The Quality of Our Nation's Waters

Nutrients and Pesticides—A Summary

STREAMS AND GROUND WATER in basins with significant agricultural or urban development, or with a mix of these land uses, almost always contain complex mixtures of nutrients (nitrogen and phosphorus compounds) and pesticides. These mixtures are composed of chemicals in current use, others that were used historically (such as DDT, which was banned in the early 1970s), and chemical breakdown products. The types and concentrations of nutrients and pesticides found in streams and ground water are closely linked to land use and the chemicals applied in each setting, such as fertilizers and pesticides applied in agricultural and urban areas, and nutrients from animal and human wastes. Thus, local and regional management of fertilizer and pesticide use can go a long way toward improving water-quality conditions.

Land and chemical use are not, however, the sole predictors of water quality. Concentrations of nutrients and pesticides vary considerably from season to season, as well as among watersheds with differing vulnerability to contamination. Natural features, such as geology and soils, and land-management practices, such as tile drainage and irrigation, can affect the movement of chemicals over land or to aquifers and can thereby exert important local and regional controls on water quality. Understanding the national, regional, and local importance of land and chemical use, natural features, and management practices on water quality increases the effectiveness of policies designed to protect water resources in diverse settings.

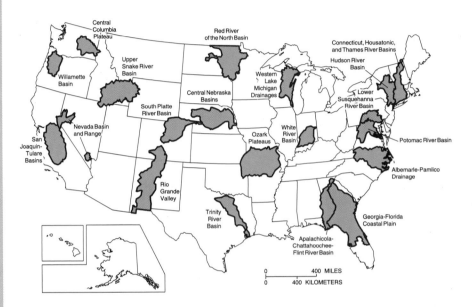

National and regional insights on nutrients and pesticides in streams and ground water are based on findings from studies completed in 1998 by the National Water-Quality Assessment (NAWQA) Program in 20 of the Nation's most important river basins and aquifer systems.

Extent of nutrient contamination and possible concerns

Nitrate, the form of nitrogen most related to human health, generally does not pose a health risk for residents whose drinking water comes from streams or from aquifers buried relatively deep beneath the land. Health risks increase in aquifers located in vulnerable geologic settings, such as in sand, gravel, or karst (weathered carbonate rock), that allow rapid movement of water. In 4 of 33 major drinking-water aquifers sampled, the U.S. Environmental Protection Agency (USEPA) drinking-water standard for nitrate was exceeded in more than 15 percent of samples collected. These aquifers, all of which underlie intensive agricultural areas, are in vulnerable geologic settings in the Central Valley of California, the Great Plains, and parts of the Mid-Atlantic region.

The most prevalent nitrate contamination was detected in shallow ground water (less than 100 feet below land surface) beneath agricultural and urban areas, where about 15 percent of all samples exceeded the USEPA drinking-water standard. This finding raises potential concerns for human health, particularly in rural agricultural areas where shallow ground water is used for domestic water supply. Furthermore, high levels of nitrate in shallow ground water may serve as an early warning of possible future contamination of older underlying ground water, which is a common source for public water supply.

Concentrations of nitrogen and phosphorus commonly exceed levels that can contribute to excessive growth of algae and other nuisance plants in streams. For example, average annual concentra-

Nutrients and pesticides and their connection to land use

Relative levels of contamination are closely linked to land use and to the amounts and types of chemicals used in each setting. Some of the highest concentrations of nitrogen and herbicides, including those most heavily used (such as atrazine, metolachlor, alachlor, and cyanazine) were detected in samples collected from streams and shallow ground water in agricultural areas. Some of the highest concentrations of phosphorus and insecticides, including those currently used (such as diazinon, carbaryl, and malathion) and those historically used (such as DDT, dieldrin, and chlordane) were detected in samples collected from urban streams.

RELATIVE LEVEL OF CONTAMINATION

Streams

	Urban areas	Agricultural areas	Undeveloped areas
Nitrogen	Medium	Medium–High	Low
Phosphorus	Medium–High	Medium–High	Low
Herbicides	Medium	Low–High	No data
Currently used insecticides	Medium-High	Low–Medium	No data
Historically used insecticides	Medium-High	Low–High	Low

Shallow Ground Water

	Urban areas	Agricultural areas
Nitrogen	Medium	High
Phosphorus	Low	Low
Herbicides	Medium	Medium–High
Currently used insecticides	Low–Medium	Low–Medium
Historically used insecticides	Low-High	Low-High

tions of total phosphorus in three-fourths of streams in urban and agricultural areas were greater than the USEPA desired goal for preventing nuisance plant growth, including algae, in streams. Such growth can clog water intake pipes and filters and can interfere with recreational activities, such as fishing, swimming, and boating. In addition, subsequent decay of algae can result in foul odors, bad taste in drinking water, and low dissolved oxygen in aquatic habitats.

Extent of pesticide contamination and possible concerns

The NAWQA Program measured 83 pesticides and breakdown products in water and 32 pesticides in fish or bed sediment. At least one pesticide was found in almost every water and fish sample collected from streams and in more than one-half of shallow wells sampled in agricultural and urban areas. Moreover, individual pesticides seldom occurred alone. Almost every sample from streams and about one-half of samples from wells with a detected pesticide contained two or more pesticides. Although pesticides frequently are found in water, their potential effects on humans and aquatic

life are not fully understood. Potential effects must be gauged from a combination of established water-quality standards and guidelines, and by careful consideration of uncertainties and the potential for unaccounted influences due to complexities related to pesticide occurrence.

The good news is that concentrations of individual pesticides in samples from wells and as annual averages in streams were almost always lower than current USEPA drinking-water standards and guidelines. Standards and guidelines have been established for 46 pesticides and breakdown products. Effects of pesticides on aquatic life, however, are a concern based on U.S. and Canadian aquatic-life guidelines, which have been established for 28 of the pesticides measured. More than one-half of agricultural and urban streams sampled had concentrations of at least one pesticide that exceeded a guideline for the protection of aquatic life.

Potential risks to humans and aquatic life implied by NAWQA pesticide findings can be only partially addressed by comparison to established standards and guidelines. Many pesticides and their breakdown products do not have standards or guidelines, and current standards and guidelines do not yet account for exposure

to mixtures and seasonal pulses of high concentrations. In addition, potential effects on reproductive, nervous, and immune systems, as well as on chemically sensitive individuals, are not yet well understood. For example, some of the most frequently detected pesticides are suspected endocrine disrupters that have potential to affect reproduction or development of aquatic organisms or wildlife by interfering with natural hormones.

The widespread occurrence of pesticides in water and the pervasive uncertainty in assessing potential effects on humans and aquatic life make pesticide contamination a particularly difficult water-quality problem to resolve. More information is needed on potential effects that are not well understood. In the meantime, our understanding of patterns of pesticide contamination in relation to land use, pesticide use, and the natural characteristics of hydrologic systems can help us to reduce the amounts of pesticides that reach streams and ground water.

Thomas L. Huntzinger

Concentrations of nutrients and pesticides in streams and shallow ground water generally increase with increasing amounts of agricultural and urban land. This pattern is evident within small watersheds, as well as regionally, where similar land-use settings and chemical applications extend over broad areas. For example, intensive herbicide and fertilizer use in the Upper Midwest has resulted in elevated concentrations of atrazine, nitrogen, and phosphorus in streams throughout the region, including the Mississippi River. Management strategies that are successful in reducing use and transport of herbicides and fertilizers could lead to regional improvements in water quality.

Importance of seasonal and geographic patterns in determining protection strategies

Seasonal patterns in water quality of streams emerged in most basins. The patterns reflect many factors, but mainly the timing and amount of chemical use, the frequency and magnitude of runoff from rainstorms or snowmelt, and specific land-management practices, such as irrigation and tile drainage. Concentrations of nutrients and pesticides are highest during runoff following chemical applications. The seasonal nature of these factors dictate the timing of elevated concentrations in drinking-water sources and aquatic habitats.

The geographic distribution of natural features (including topography, geology, soils, hydrology, and climate) and land-management practices (including tile drainage, irrigation, and conservation strategies) also affect the occurrence of nutrients and pesticides in water. These factors make some areas more vulnerable to contamination than other areas, thus, concentrations of nutrients and pesticides can vary among seemingly similar land uses and types of chemical applications.

Ground water is most vulnerable to contamination in well-drained areas with permeable soils that are underlain by sand and gravel or karst. Examples are the Platte River Valley in Colorado and Nebraska, and karst regions within the Susquehanna and Potomac River Basins in Pennsylvania, Maryland, and Virginia. In contrast, streams are most vulnerable in basins with poorly drained clay soils, steep slopes, or limited vegetation to slow runoff. Tile drains and ditches also provide

quick pathways for nutrient and pesticide runoff to streams, such as in the White River Basin in Indiana.

Patterns in regional vulnerability are evident where similar natural features, land use, and land-management practices extend over broad areas. For example, ground water underlying intensive agriculture in parts of the Upper Midwest is minimally contaminated where it is protected by relatively impermeable soils and glacial till that cover much of the region. Local hotspots of nitrate and pesticide contamination occur in the region where ancient glacial streams

deposited sand and gravel, which enable rapid infiltration and downward movement of water and chemicals. Another example is in the Southeast, where streams and ground water contain relatively low concentrations of nitrogen, partly because soil and hydrologic characteristics in this region favor conversion to nitrogen gas. In contrast, relatively high nitrogen concentrations occur in streams and shallow ground water in the Central Valley of California and parts of the Northwest, Great Plains, and Mid-Atlantic regions, because natural characteristics favor transport of nitrogen.

Peter A. Steeves

Concentrations of nutrients and pesticides generally are higher and more prevalent in streams than in ground water; however, indications of emerging ground-water contamination are important because ground-water contamination is difficult to reverse. Ground-water flow rates are slow, and a contaminated aquifer can take years or even decades to recover.

Water-quality changes

Water quality is constantly changing, from season to season and from year to year. Long-term trends, as captured by the question "Are things getting better or worse?," are sometimes difficult to distinguish from short-term fluctuations. For many chemicals, it is too early to tell whether conditions are better or worse, because historical data are insufficient or too inconsistent to measure trends. Despite these challenges, some trends are evident from monitoring of pesticides and nutrients. These trends show that changes in water quality over time frequently are controlled by factors similar to those that affect geographic variability, including natural features, chemical use, and management practices.

One of the most striking trends is a national reduction in concentrations of organochlorine insecticides, such as DDT, dieldrin, and chlordane, in whole fish. Concentrations of DDT also have decreased in sediment, as indicated in sediment-core samples from urban and agricultural reservoirs and lakes. Just as notable as these declines, however, is that these persistent insecticides

Terry R. Maret

still are found at elevated levels in fish and streambed sediment in many urban and agricultural streams across the Nation.

Historical data also show that total nitrogen concentrations have remained stable over the past 20 years in rivers downstream from wastewater treatment plants, such as in the Trinity River in Dallas, Texas. Improved treatment has resulted in decreased concentrations of ammonia and phosphorus despite urban population growth, but has also resulted in changes in the forms of nitrogen in the river. Ammonia is converted to nitrate, which makes the

Kevin F. Dennehy

discharge less toxic to fish but may not resolve problems with excessive plant growth.

Changes in concentrations of modern, short-lived pesticides follow changes in use, often focused in specific regions and land-use areas. For example, increases in acetochlor and decreases in alachlor are evident in some streams in the Upper Midwest, where acetochlor partially replaced alachlor for control of weeds in corn and soybeans beginning in 1994. The changes in use are reflected quickly in stream quality, generally within 1 to 2 years.

In contrast, ground-water quality responds more slowly to changes in chemical use or land-management practices, typically lagging by many years and even decades. Local variations in natural features, such as soil types and amounts of recharge, can result in variable rates of ground-water flow, which thereby affect the long-term response to land practices. For example, concentrations of nitrate decreased significantly (from about 18 milligrams per liter in the mid-1980s to less than 2 milligrams per liter in the mid-1990s) in ground water underlying parts of the Central Platte Natural Resources District, Nebraska, after implementation of fertilizer management strategies. Yet, despite implementation of the strategies, the response has been delayed in other parts of the District because of differences in local features controlling ground-water flow. Specifically, concentrations of nitrate remained greater than two times the USEPA drinking-water standard in nearly one-fourth of wells in one area sampled by the District in the mid-1990s.

Science-based lessons for water-quality management and policy

Reductions of nutrient and (or) pesticide concentrations in streams and ground water clearly require management strategies that focus on reducing chemical use and subsequent transport in the hydrologic system. For these strategies to be effective, they should be developed with careful consideration of the patterns and complexities of contaminant occurrence, behavior, and influences on water quality. NAWQA results indicate four basic considerations that are critical for managing and protecting water resources in diverse settings across the Nation.

First, local and regional management strategies are needed to account for geographic patterns in land use, chemical use, and natural factors, which govern hydrologic behavior and vulnerability to contamination. Second, nutrients and pesticides are readily transported among surface water, ground water, and the atmosphere and, therefore, environmental policies that simultaneously address the entire hydrologic system are needed to protect water quality. Third, a top priority should be to reduce the uncertainty in estimates of the risks of pesticides and other contaminants to humans and aquatic life. This will require improved information on the nature of exposure and effects, and development of standards, guidelines, and monitoring programs that address the many complexities in contaminant occurrence. For example, neither current standards and guidelines nor associated monitoring programs, particularly with regard to pesticides, account for contamination that occurs as mixtures of various parent compounds and degradation products, or that is characterized by lengthy periods of low concentrations punctuated by brief, seasonal periods of higher concentrations. Finally, continued development of reliable predictive models is an essential element of cost-effective strategies to anticipate and manage nutrient and pesticide concentrations over a wide range of possible circumstances, over broad regions, and for the long term. An understanding of these considerations will help water managers and policy makers in their implementation of environmental control and protection strategies, in investments in monitoring and science, and in the development of future environmental policies, standards, and guidelines.

This Fact Sheet accompanies a USGS nontechnical publication titled "The Quality of Our Nation's Waters— Nutrients and Pesticides" (USGS Circular 1225). These publications are available through the NAWQA Program. For information, contact:

Chief, NAWQA Program
U.S. Geological Survey
413 National Center
Reston, VA 20192

phone: (703) 648-5716
fax: (703) 648-6693
email: nawqa_whq@usgs.gov
http://water.usgs.gov/lookup/get?nawqa

U.S. Department of the Interior
U.S. Geological Survey

The Quality of Our Nation's Waters

Nutrients and Pesticides

Circular 1225

U.S. Department of the Interior
Bruce Babbitt, Secretary

U.S. Geological Survey
Charles G. Groat, Director

Reston, Virginia, 1999

For additional copies please contact:

USGS Information Services
Box 25286
Denver CO 80225

For more information about the USGS and its products:

Telephone: 1-800-USA-MAPS
World Wide Web: http://www.usgs.gov

Suggested citation for this report:

U.S. Geological Survey, 1999, The Quality of Our Nation's Waters—
Nutrients and Pesticides: U.S. Geological Survey Circular 1225, 82 p.

Library of Congress Cataloging-in-Publication Data
The quality of our nation's waters : nutrients and pesticides/
    Gregory J. Fuhrer... [et al.].
        p. cm.— (U.S. Geological Survey circular; 1225)
    Includes bibliographic references.
    ISBN 0-607-92296-6
    1. Nutrient pollution of water—United States. 2. Pesticides-
-Environmental aspects—United States. 3. Water quality—United
States. I. Fuhrer, Gregory J. II Series.
TD427.N87Q33 1999
363.739'42'0973—dc21

99—26892
          CIP

National Water-Quality Assessment Program

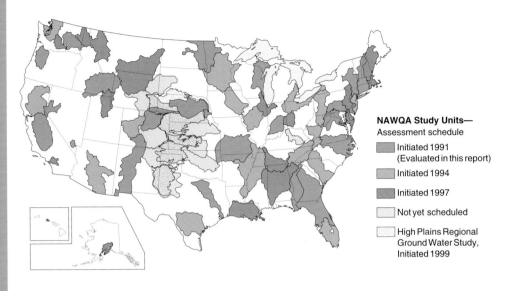

NAWQA Study Units—
Assessment schedule

- Initiated 1991 (Evaluated in this report)
- Initiated 1994
- Initiated 1997
- Not yet scheduled
- High Plains Regional Ground Water Study, Initiated 1999

"The Nation's water resources are the basis for life and our economic vitality. These resources support a complex web of human activities and fishery and wildlife needs that depend upon clean water. Demands for good quality water for drinking, recreation, farming, and industry are rising, and as a result, the American public is concerned about the condition and sustainability of our water resources. As part of the National Water-Quality Assessment Program, the U.S. Geological Survey will continue to work with other Federal, State, and local agencies to better understand how natural and human influences affect water quality in different parts of the Nation. Without this understanding, we can not wisely manage these resources."

Bruce Babbitt, Secretary
U.S. Department of the Interior

In 1991, the U.S. Congress appropriated funds for the U.S. Geological Survey (USGS) to begin the National Water-Quality Assessment (NAWQA) Program. As part of the NAWQA Program, the USGS works with other Federal, State, and local agencies to understand the spatial extent of water quality, how water quality changes with time, and how human activities and natural factors affect water quality across the Nation. Such understanding can help resource managers and policy makers to better anticipate, prioritize, and manage water quality in different hydrologic and land-use settings and to consider key natural processes and human factors in resource strategies and policies designed to restore and protect water quality.

The NAWQA Program focuses on water quality in more than 50 major river basins and aquifer systems. Together, these include water resources available to more than 60 percent of the population in watersheds that cover about one-half of the land area of the conterminous United States. NAWQA began investigations in 20 of these areas in 1991 and phased in work in more than 30 additional basins by 1997. Investigations in these basins, referred to as "Study Units," use a nationally consistent scientific approach and standardized methods. The consistent design facilitates investigations of local conditions and trends within individual Study Units, while also providing a basis to make comparisons among Study Units. The comparisons demonstrate that water-quality patterns are related to chemical use, land use, climate, geology, topography, and soils, and thereby improve our understanding of how and why water quality varies regionally and nationally.

Introduction to this report and the NAWQA series
The Quality of Our Nation's Waters

Douglas A. Harned

This report is the first in a series of nontechnical publications, *The Quality of Our Nation's Waters*, that describe major findings of the NAWQA Program on water-quality issues of regional and national concern. This first report presents insights on nutrients and pesticides in water and on pesticides in bed sediment and fish tissue. It represents a compilation of findings in the first 20 Study Units.[1] Subsequent reports in this series will cover other water-quality constituents of concern, such as radon, arsenic, other trace elements, and industrial chemicals, as well as physical and chemical effects on aquatic ecosystems. Each report will build toward a more comprehensive understanding of regional and national water resources as assessments in other Study Units are completed and as scientific models and tools that link water-quality conditions, dominant sources, and environmental characteristics are applied in geographic areas not covered by NAWQA Study Units.

The information in this series is intended primarily for those interested or involved in resource management, conservation, regulation, and policy making at regional and national levels. In addition, the information might interest those at a local level who simply wish to know more about the general quality of streams and ground water in areas near where they live, and how that quality compares to other areas across the Nation.

Charles G. Groat, Director
U.S. Geological Survey

---

[1] Summaries of water-quality assessments for the first 20 Study Units are available as USGS Circulars and on the World Wide Web. Information on accessing these summaries is provided on p. 80.

Contents

The first two sections provide a general overview of findings on nutrients and pesticides and their implications for water-resource management and protection. More detailed technical discussions of the sources, distributions, and potential effects of these chemicals are provided in subsequent sections.

Societal concerns for the quality of our water resources continue, as many of the Nation's streams and coastal waters do not meet water-quality goals. States report that 40 percent of the waters they surveyed are too contaminated for basic uses, such as fishing and swimming. Some progress has been made since passage of the Clean Water Act in 1972. Since the early 1970s, private and public sectors have spent more than $500 billion on water-pollution control, much of which has been directed toward municipal and industrial point sources.[1] Although some violations still occur, this legislation has had a positive effect on limiting contaminants from point sources entering streams.

Progress in cleaning up contamination from point sources has not yet been matched by control of contaminated runoff from nonpoint sources, including fertilizers and pesticides applied in agricultural and urban areas, and nutrients from human and animal wastes. The challenges are great because nonpoint sources are ubiquitous yet highly variable causes of water-quality problems, making them difficult to evaluate and control.

Beginning in the early 1990s, widespread environmental and public-health concerns resulted in a Federal water-quality initiative to work with the Nation's farmers to protect surface water and ground water from nutrient and pesticide contamination. To address these national concerns, nutrients and pesticides were two of the first water-quality issues evaluated by the NAWQA Program. This report, which presents regional and national insights on these chemicals, is based on a compilation of findings from the first 20 NAWQA Study Units.

Concerns about nutrients

Nitrogen and phosphorus are essential for healthy plant and animal populations; however, elevated concentrations of these nutrients can degrade water quality. Excessive nitrate in drinking water can result in "blue-baby syndrome," which causes oxygen levels in the blood of infants to be low, sometimes fatally. Elevated nitrogen and phosphorus concentrations in surface water can trigger eutrophication, resulting in excessive, often unsightly, growth of algae and other nuisance aquatic plants. These plants can clog water intake pipes and filters and can interfere with recreational activities, such as fishing, swimming, and boating. Subsequent decay of algae can result in foul odors, bad taste, and low dissolved oxygen in water (hypoxia). Excessive nutrient concentrations have been linked to hypoxic conditions, such as those found in the Gulf of Mexico, which can harm fish and shellfish that are economically and ecologically important to the Nation. High nutrient concentrations also are believed to be one cause for the growth of the dinoflagellate *Pfiesteria*, found in Atlantic coastal waters. This form of algae is potentially toxic to fish and other organisms, including humans.

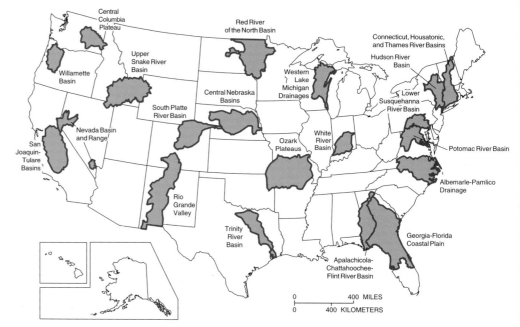

One of the challenges and goals for NAWQA in the first 20 Study Units was to explain where nutrients and pesticides (which include herbicides, insecticides, and other classes of pesticides) commonly occur, and why some land-use and environmental settings are more vulnerable to contamination than others, particularly during certain times of the year. Stream quality was monitored seasonally and during high-flow events, such as storms and periods of peak irrigation, as well as over several years, to better understand when changes occur. By NAWQA design, an initial 3 to 4 years of intensive study are followed by 6 to 7 years of low-level monitoring, at which time intensive study resumes to assess water-quality changes.

Streams and shallow ground water in agricultural, urban, and some undeveloped (mostly forested) settings were studied in the first 20 Study Units. The agricultural areas are diverse in climate and geography, and they span coastal, desert, and temperate environmental settings. They include areas of corn and soybean production in the Midwest; areas of production of wheat and other grains in the Great Plains; areas of mixed row crop and poultry production in the East; rangeland grazing and cattle feeding operations in the arid Southwest; and areas of intensive production of grain, fruits and nuts, vegetables, and specialty crops in California and the Pacific Northwest.

Sampling of streams and shallow ground water in urban areas represented primarily residential land use, typically with low to medium population densities.[2] In general, the urban assessments focused on nonpoint sources of contaminants, although some sampling of rivers was done downstream from major metropolitan areas (such as Atlanta and Denver, which have point discharges from municipal wastewater treatment plants).

Nutrients and pesticides also were assessed in major rivers and in aquifers commonly used for drinking water. These resources represent integrated water-quality effects from multiple land uses and environmental settings that occur within relatively large contributing areas.

Concerns about pesticides

Pesticides, used to control weeds, insects, and other pests, receive widespread public attention because of potential impacts on humans and the environment. Depending on the chemical, possible health effects from overexposure to pesticides include cancer, reproductive or nervous-system disorders, and acute toxicity. Similar effects are possible in the aquatic environment. Recent studies suggest that some pesticides can disrupt endocrine systems and affect reproduction by interfering with natural hormones.

The NAWQA Program was not intended to assess the quality of the Nation's drinking water, such as by monitoring water from taps. Rather, NAWQA assessments focus on the quality of the resource itself, thereby complementing many ongoing Federal, State, and local drinking-water monitoring programs. Comparisons made in this report to drinking-water standards and guidelines are made only in the context of the available untreated resource.

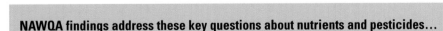

NAWQA findings address these key questions about nutrients and pesticides...

• Which nutrients and pesticides are found in streams and ground water across the Nation? At what concentrations?

• Are elevated concentrations more prevalent in certain geographic regions and environmental settings?

• How do differences in land use, chemical use, land-management practices, and natural processes help explain differences in vulnerability to contamination of streams and ground water?

• Are nutrient and pesticide concentrations elevated only at certain times of the year? Are concentrations changing over time?

• What are the implications to human health and the environment?

• How is this information useful for guiding future research, monitoring, and water-management and protection strategies related to nutrients and pesticides?

NAWQA

findings focus on how, when, and why nutrients and pesticides vary across the Nation. This information is useful to help anticipate, prioritize, and manage water-quality conditions in different land uses and environmental settings. In addition, the findings point to several science-based considerations for policies and strategies designed to restore and protect the quality of our most vulnerable waters.

Do NAWQA findings substantiate national concerns?

NAWQA findings indicate that streams and ground water in basins with significant agricultural or urban development, or with a mix of these land uses, almost always contain complex mixtures of nutrients and pesticides. Concentrations of nitrogen and phosphorus commonly exceed levels that can contribute to excessive plant growth in streams. For example, average annual concentrations of phosphorus in three-fourths of streams in urban and agricultural areas were greater than the U.S. Environmental Protection Agency (USEPA) desired goal for preventing nuisance plant growth in streams. Nitrate generally does not pose a health risk for residents whose drinking water comes from streams or from aquifers buried relatively deep beneath the land. Health risks increase in those aquifers located in geologic settings, such as in sand, gravel, or karst (weathered carbonate rock), that enable rapid movement of water. The most prevalent nitrate contamination was detected in shallow ground water (less than 100 feet below land surface) beneath agricultural and urban areas. This finding raises potential concerns for human health, particularly in rural agricultural areas where shallow ground water is used for domestic water supply. Furthermore, high levels of nitrate in shallow ground water may serve as an early warning of possible future contamination of older underlying ground water, which is commonly a primary source for public water supply.

At least one pesticide was found in almost every water and fish sample collected from streams and in more than one-half of shallow wells sampled in agricultural and urban areas. Moreover, individual pesticides seldom occurred alone. Almost every sample from streams and about one-half of samples from wells with a detected pesticide contained two or more pesticides. Concentrations of individual pesticides in samples from wells and as annual averages in streams were almost always lower than current USEPA drinking-water standards and guidelines. Standards and guidelines have been established for 46 of the 83 pesticides and breakdown products measured in water. Effects of pesticides on aquatic life, however, are a concern based on U.S. and Canadian guidelines, which have been established for 28 of the pesticides measured. More than one-half of agricultural and urban streams sampled had concentrations of at least one pesticide that exceeded a guideline for the protection of aquatic life.

Potential risks to humans and aquatic life implied by NAWQA pesticide findings can be only partially addressed by comparison to established standards and guidelines. Many pesticides and their breakdown

Nutrients and pesticides and their connection to land use

Relative levels of contamination are closely linked to land use and to the amounts and types of chemicals used in each setting. Thus, local and regional management of chemical use can go a long way toward improving water-quality conditions.

RELATIVE LEVEL OF CONTAMINATION

	Streams				Shallow Ground Water	
	Urban areas	Agricultural areas	Undeveloped areas		Urban areas	Agricultural areas
Nitrogen	Medium	Medium–High	Low	Nitrogen	Medium	High
Phosphorus	Medium–High	Medium–High	Low	Phosphorus	Low	Low
Herbicides	Medium	Low–High	No data	Herbicides	Medium	Medium–High
Currently used insecticides	Medium-High	Low–Medium	No data	Currently used insecticides	Low–Medium	Low–Medium
Historically used insecticides	Medium-High	Low–High	Low	Historically used insecticides	Low–High	Low–High

products do not have standards or guidelines, and current standards and guidelines do not yet account for exposure to mixtures and seasonal pulses of high concentrations. In addition, potential effects on reproductive, nervous, and immune systems, as well as on chemically sensitive individuals, are not yet well understood. For example, some of the most frequently detected pesticides are suspected endocrine disrupters that have potential to affect reproduction or development of aquatic organisms or wildlife by interfering with natural hormones.

Are seasonal and geographic patterns evident and important in determining protection strategies?

Land and chemical use are not the sole predictors of water quality. Concentrations of nutrients and pesticides vary considerably from season to season, as well as among watersheds with differing vulnerability to contamination. Natural features, such as geology and soils, and land-management practices, such as tile drainage and irrigation, can affect the movement of chemicals over land or to aquifers and can thereby exert local and regional controls on water quality. Understanding the national, regional, and local importance of land and chemical use, natural features, and management practices on water quality increases the effectiveness of policies designed to protect water resources in diverse settings.

Seasonal patterns in water quality of streams emerged in most basins. The patterns reflect many factors, but mainly the timing and amount of chemical use, the frequency and magnitude of runoff from rainstorms or snowmelt, and specific land-management practices, such as irrigation and tile drainage. Concentrations of nutrients and pesticides are highest during runoff following chemical applications. The seasonal nature of these factors dictate the timing of elevated concentrations in drinking-water sources and aquatic habitats.

Natural features and land-management practices make some areas more vulnerable to contamination than other areas, thus, concentrations of nutrients and pesticides can vary among seemingly similar land uses and types of chemical applications. Patterns are most evident on a local scale, but they also occur regionally where similar natural features, land use, and land-management practices extend over broad areas. For example, ground water underlying intensive agriculture in parts of the Upper Midwest is minimally contaminated where it is protected by relatively impermeable soils and glacial till that cover much of the region. Tile drains and ditches commonly provide quick pathways for nutrient and pesticide runoff to streams in this area. Another example is in the Southeast, where streams and ground water commonly contain relatively low concentrations of nitrogen, partly because soil and hydrologic characteristics in this region favor conversion to nitrogen gas. In contrast, relatively high nitrogen concentrations occur in streams and shallow ground water in the Central Valley of California and parts of the Northwest, Great Plains, and Mid-Atlantic regions because natural characteristics favor transport of nitrogen.

Is water quality getting better or worse?

Water quality is constantly changing, from season to season and from year to year. Long-term trends are sometimes difficult to distinguish from short-term fluctuations. For many chemicals, it is too early to tell whether conditions are getting better or worse because historical data are insufficient or too inconsistent to measure trends. Despite these challenges, some trends are evident from monitoring of nutrients and pesticides. These trends show that changes in water quality over time frequently are controlled by factors similar to those that affect geographic variability, including natural features, chemical use, and management practices. For example, concentrations of the organochlorine insecticide DDT have decreased in sediment and fish since restrictions were imposed on its production and distribution in the 1970s.

Changes in concentrations of modern, short-lived pesticides also follow changes in use; these changes are often focused in specific regions and land-use areas. For example, increases in acetochlor and decreases in alachlor are evident in some streams in the Upper Midwest, where acetochlor partially replaced alachlor for control of weeds in corn and soybeans beginning in 1994. The changes in use are reflected quickly in stream quality, generally within 1 to 2 years. In contrast, ground water responds more slowly to changes in chemical use or land-management practices because of slower travel-times. This response can be delayed by years or decades.

Water-quality patterns...

Transport of a chemical compound in the environment depends on its mobility. Some compounds, such as nitrate and atrazine, readily dissolve and move with water in both streams and ground water. Many forms of phosphorus, however, attach to soil particles rather than dissolve; a large proportion of such compounds is transported to streams with eroded soil, particularly during times of high runoff from precipitation or irrigation. Ground water typically is not vulnerable to contamination by compounds that attach to soils.

The transport of a chemical compound in the environment also depends on its persistence. Some pesticides are not readily broken down by microorganisms or other processes in the natural environment. For example, DDT and chlordane can persist in soil, water, sediment, and animal tissue for years and even decades. Other pesticides, such as carbaryl, are relatively unstable in water and break down to other compounds in days or weeks. Chemical compounds that persist for a long time are likely to be transported farther than compounds that are short-lived.

Some of the highest levels of nitrogen occur in streams and ground water in agricultural areas

Applications of fertilizers, manure, and pesticides have degraded the quality of streams and shallow ground water in agricultural areas and have resulted in some of the highest concentrations of nitrogen measured in NAWQA studies. Concentrations of nitrogen in nearly half of the streams sampled in agricultural areas ranked among the highest of all streams measured in the first 20 Study Units. Concentrations of nitrate exceeded the USEPA drinking-water standard of 10 milligrams per liter (as nitrogen) in 15 percent of samples collected in shallow ground water beneath agricultural and urban land, signifying a possible concern in some rural areas where shallow aquifers are used for drinking-water supply.

Phosphorus is elevated, too

Compared to nitrogen, a smaller proportion of phosphorus (originating mostly from livestock wastes or fertilizers) was lost from watersheds to streams. The annual amounts of total phosphorus and total nitrogen measured in agricultural streams were equivalent to less than 20 percent of the phosphorus and less than 50 percent of the nitrogen that was applied annually to the land. This is consistent with the general tendency of phosphorus to attach to soil particles and move with runoff to surface water. Even with the lower losses from land for phosphorus than for nitrogen, however, phosphorus is more likely to reach concentrations that can cause excessive aquatic plant growth. Nitrogen concentrations are rarely low enough to limit aquatic plant growth in freshwater, whereas phosphorus concentrations can be low enough to limit such growth. Hence, excessive aquatic plant growth and eutrophication in freshwater generally result from elevated phosphorus concentrations (typically greater than 0.1 milligram per liter). In contrast, nitrogen is typically the limiting nutrient for aquatic plant growth in saltwater and coastal waters.

Pesticides—primarily herbicides—are found frequently in agricultural streams and shallow ground water

Extensive herbicide use in agricultural areas (accounting for about 70 percent of total national use of pesticides) has resulted in widespread occurrence of herbicides in agricultural streams and shallow ground water. The highest rates of detection for the most heavily used herbicides—atrazine, metolachlor, alachlor, and cyanazine—were found in streams and shallow ground water in agricultural areas. Insecticides were frequently detected in some streams draining watersheds with high insecticide use but were less frequently detected in shallow ground water because most insecticides are applied at lower levels than herbicides and tend to sorb onto soil or degrade quickly after application.

…in agricultural areas

Health effects of pesticides are not adequately understood

Concentrations of individual pesticides generally were low compared to USEPA drinking-water standards and guidelines; pesticides exceeded standards or guidelines in less than 1 percent of sampled wells. This good news, however, is tempered by the current uncertainty in estimating risks of pesticide exposure. For example, most contamination occurred as pesticide mixtures, such as atrazine, metolachlor, and other pesticides, whereas most toxicity and exposure assessments are based on controlled experiments with a single contaminant. In addition, some breakdown products, for which there are no established standards or guidelines, may have effects similar to their parent pesticides. Finally, water-quality standards and guidelines have been established for only about one-half of the pesticides measured in NAWQA water samples.

Aquatic life may be at more risk than human health

Effects on aquatic organisms may be greater than on humans in many agricultural areas. Although there are no USEPA aquatic-life criteria for the major herbicides, Canadian guidelines were exceeded at 17 of the 40 agricultural streams studied, most commonly for atrazine or cyanazine. Also, currently used insecticides exceeded guidelines for aquatic life in at least one water sample from 18 of the 40 agricultural streams. The major organochlorine insecticides, such as DDT, dieldrin, and chlordane (which no longer are used but remain widely detected in sediment and fish in agricultural streams) exceeded recommended sediment-quality guidelines for protection of aquatic life at about 15 percent of agricultural sites.

Jana S. Stewart

Jeffrey D. Stoner

Larry F. Land

Water-quality patterns...

Insecticides typically were detected in urban areas, sometimes at high concentrations

Urban areas, covering less than 5 percent of land in the continental United States, traditionally have not been recognized as important contributors to pesticide contamination, especially when compared to agricultural land, which covers more than 50 percent of the United States. Findings in the first 20 Study Units, however, show a widespread occurrence of some insecticides commonly used around homes and gardens and in commercial and public areas. In fact, these insecticides occurred at higher frequencies, and usually at higher concentrations, in urban streams than in agricultural streams. Most common were diazinon, carbaryl, chlorpyrifos, and malathion. As in agricultural areas, insecticides were detected in ground water less frequently than in streams. Some herbicides—including atrazine, simazine, and prometon, which are used to control weeds in lawns and golf courses, and along roads and rights-of-way— also occurred frequently in samples collected from streams and shallow ground water in urban areas.

Concentrations of insecticides in urban streams commonly exceeded guidelines for protection of aquatic life

Insecticides, which generally are more toxic to aquatic life than herbicides, frequently exceeded USEPA, Canadian, or International Joint Commission water-quality guidelines in urban streams. Almost every urban stream sampled had concentrations of insecticides that exceeded at least one guideline, and most had concentrations that exceeded a guideline in 10 to 40 percent of samples collected throughout the year.

Urban streams had the highest frequencies of occurrence of DDT, chlordane, and dieldrin in fish and sediment, and the highest concentrations of chlordane and dieldrin

DDT is an insecticide that commonly was used in the United States until the early 1970s to control insects on cropland and lawns and mosquitoes in populated areas. Chlordane and aldrin (the parent compound that breaks down to dieldrin) were used widely until the late 1980s to control termites. Since the use of DDT was restricted, concentrations have decreased in sediment in urban areas, as indicated by sediment-core samples from urban reservoirs and lakes. Similar declines are not yet evident in concentrations of chlordane and dieldrin in sediment, most likely because of their continued use into the late 1980s. Despite downward trends in some areas, organochlorine insecticides commonly are found at elevated levels in bed sediment and fish in urban streams. Sediment-quality guidelines for protection of aquatic life were exceeded at nearly 40

...in urban areas

percent of urban sites, and concentrations in whole fish exceeded guidelines for protection of wildlife at 20 percent of urban sites. Although most urban streams are not used for drinking water, the frequent occurrence of insecticides in water, sediment, and fish is a potential concern for recreational use and for fish consumption.

Complex mixtures of pesticides commonly occur in urban streams

Similar to agricultural pesticides, urban pesticides commonly occurred in mixtures. More than 10 percent of urban stream samples contained a mixture of the insecticides diazinon and chlorpyrifos, along with at least four herbicides. Two of the most common herbicides in these mixtures were simazine and prometon.

Concentrations of phosphorus were elevated in urban streams

Concentrations of total phosphorus in streams generally were higher in urban areas than in agricultural areas; concentrations commonly exceeded the USEPA desired goal (0.1 milligram per liter) to control excessive growth of algae and other nuisance plants in streams. Elevated concentrations of phosphorus are, in part, due to effluent from wastewater treatment plants, despite some long-term decreases in phosphorus resulting from improved treatment technology. The highest concentrations of total phosphorus were in streams in semiarid western and southwestern cities, where discharges from wastewater treatment plants may account for a significant proportion of streamflow. Concentrations of phosphorus also were high in urban areas in the East.

Kevin F. Dennehy

Nitrogen levels have remained nearly unchanged in rivers downstream from wastewater treatment plants

Although NAWQA focused mostly on nonpoint sources of nutrients, sampling of some rivers downstream from wastewater treatment plants showed that total nitrogen levels have remained nearly stable since the 1970s. Improvements in wastewater treatment have kept pace with urban population growth in major metropolitan areas. However, wastewater treatment has resulted in changes in the forms of nitrogen in the water; specifically, nitrogen in the form of ammonia commonly is converted to nitrate during the treatment process. The conversion makes the discharge less toxic to fish, but it may not help to resolve problems with excessive growth of algae.

Water quality patterns...

Contamination of major aquifers is largely controlled by hydrology and land use

Concentrations of nutrients and pesticides in 33 major aquifers generally were lower than those in shallow ground water underlying agricultural and urban areas. Water that replenishes the major aquifers is derived from a variety of sources and land-use settings, and includes high-quality water from undeveloped lands. In addition, deeper aquifers generally are more protected than shallow ground water by relatively impermeable materials. Contaminants are most prevalent in major aquifers located in vulnerable geologic settings that allow rapid vertical movement of water from the shallow ground-water system. For example, in 4 of 33 major drinking-water aquifers sampled, the USEPA drinking-water standard for nitrate was exceeded in more than 15 percent of samples collected. All four aquifers are relatively shallow, in agricultural areas, and composed of sand and gravel that is vulnerable to contamination by land application of fertilizers. Water in one-third of wells sampled in major aquifers contained one or more pesticides, but only one well had a pesticide (atrazine) concentration that exceeded a drinking-water standard.

Hydrology and land use also are major factors controlling nutrient and pesticide concentrations in major rivers

Concentrations of nutrients and pesticides in major rivers reflect the proportion of urban and agricultural land in the drainage basin. River basins with large proportions of agricultural and (or) urban land had concentrations of nutrients and pesticides that were similar to those in smaller agricultural and urban streams. The greatest variety of pesticides occurred in basins draining both agricultural and urban land. Concentrations of nutrients and pesticides were moderate in major rivers draining mixed land uses because of dilution by water from undeveloped areas. None of the major rivers exceeded drinking-water standards or guidelines, although the consistent presence of pesticide mixtures remains a concern. Guidelines for the protection of aquatic life were exceeded in water at 36 percent of river sites sampled for currently used pesticides. Sediment-quality guidelines were exceeded at 11 percent of sites for DDT and other historically used insecticides, whereas concentrations of these compounds in whole fish exceeded guidelines for the protection of fish-eating wildlife at 24 percent of sites.

Key factors include soils and slope of land

Key factors governing vulnerability of surface water to contamination include the type of soil and slope of the land, both of which help to control the amount and timing runoff. Streams in basins with poorly drained clayey soils, steep slopes, and sparse vegetation generally are most vulnerable to contamination.

> Concentrations of nutrients and pesticides generally are higher and more prevalent in streams than in ground water; however, indications of emerging ground-water contamination are important because ground-water contamination is difficult to reverse. Ground-water flow rates are slow, and a contaminated aquifer can take years or even decades to recover.

…in areas with mixed land use and a range of hydrologic and environmental settings

Tile drains and urban pavement also accelerate flow to streams. In contrast, shallow ground water is most vulnerable to contamination in well-drained areas with rapid infiltration and highly permeable subsurface materials. Crop-management practices, which commonly are designed to reduce or slow the movement of sediment, nutrients, and pesticides to streams, also can increase infiltration of water and contaminants into the ground.

INCREASING POTENTIAL for nitrogen and phosphorus to enter streams…[1]

High rainfall, snowmelt, and (or) excessive irrigation, especially following recent fertilizer application

Steeply sloping areas with insufficient vegetation to slow runoff and sediment, or flat areas with artificial drains and ditches, which provide quick pathways for runoff to streams

Clayey and compacted soils underlain by poorly drained sediment and (or) nonporous bedrock, or extensive urban pavement, all of which create relatively impermeable surfaces for runoff

INCREASING POTENTIAL for nitrate to enter ground water…

High rainfall, snowmelt, and (or) excessive irrigation, especially following recent fertilizer application

Well-drained and permeable soils that are underlain by sand and gravel or karst, which enable rapid downward movement of water

Areas where crop-management practices slow runoff and allow more time for water to infiltrate into the ground

Low organic matter content and high levels of dissolved oxygen, which can minimize chemical transformations of nitrate to other forms

David F. Usher

David F. Usher

---

[1] These findings are based on general reviews of nutrient studies in agricultural and urban areas and do not necessarily indicate influences on specific forms of nitrogen or phosphorus.

James C. Petersen

The Missouri Department of Natural Resources, Division of Environmental Quality, has incorporated NAWQA stream-quality data into their database for statewide 305 (b) water-quality standards compliance monitoring. The Division will use these data to identify and prioritize problems, direct management, and assist in natural resource management, including the development of Total Maximum Daily Loads (TMDLs).

Specific science-based considerations in this section are organized into four categories:

1 Local and regional management strategies are needed to account for geographic patterns in land use, chemical use, and natural factors.

2 Development of environmental policies must consider the entire hydrologic system and its complexities, including surface-water/ground-water interactions and atmospheric contributions.

3 Water-quality standards, guidelines, and monitoring programs should reflect environmental conditions, including seasonal variations and contaminant mixtures.

4 Reliable predictive models are required to cost effectively estimate water-quality conditions that can not be directly measured for a wide range of possible circumstances.

NAWQA FINDINGS on nutrients and pesticides suggest key science-based considerations for policies and strategies designed to restore and protect water quality.

Reductions of nutrient and (or) pesticide concentrations in streams and ground water clearly require management strategies that focus on reducing chemical use and subsequent transport in the hydrologic system. For these strategies to be effective, they should be developed with careful consideration of the patterns and complexities of contaminant occurrence, behavior, and influences on water quality. For example, concentrations of nutrients and pesticides vary from season to season, as well as among watersheds with differing vulnerability to contamination. These, and other patterns and complexities, frame four basic considerations that are critical for managing and protecting water resources in diverse settings across the Nation.

First, local and regional management strategies are needed to account for geographic patterns in land use, chemical use, and natural factors, which govern hydrologic behavior and vulnerability to contamination. Second, nutrients and pesticides are readily transported among surface water, ground water, and the atmosphere and, therefore, environmental policies that simultaneously address the entire hydrologic system are needed to protect water quality. Third, a top priority should be to reduce the uncertainty in estimates of the risks of pesticides and other contaminants to humans and aquatic life. This will require improved information on the nature of exposure and effects, and development of standards, guidelines, and monitoring programs that address the many complexities in contaminant occurrence. For example, neither current standards and guidelines nor associated monitoring programs, particularly with regard to pesticides, account for contamination that occurs as mixtures of various parent compounds and degradation products, or that is characterized by lengthy periods of low concentrations punctuated by brief, seasonal periods of higher concentrations. Finally, continued development of reliable predictive models is an essential element of cost-effective strategies to anticipate and manage nutrient and pesticide concentrations over a wide range of possible circumstances, over broad regions, and for the long term.

An understanding of these considerations will help water managers and policy makers in their implementation of environmental control and protection strategies, in investments in monitoring and science, and in the development of future environmental policies, standards, and guidelines. Such information should help guide answers to frequently asked questions, such as the following: How can we prioritize assessments and monitoring of nutrients and pesticides? What should we consider in the development of source-water protection programs and Total Maximum Daily Loads (TMDLs)? How often should we monitor nutrients and pesticides? Are certain times of year more critical than other times? How much and when does ground water contribute to streams?

① Local and regional management strategies are needed to account for geographic patterns in land use, chemical use, and natural factors.

The Pennsylvania Department of Agriculture, in developing its State Pesticides and Ground-Water Strategy, has decided to prioritize ground-water areas for assessments of pesticides on the basis of NAWQA vulnerability concepts, pesticide analyses, and quality-assurance protocols.

Albert E. Becher

NAWQA data and activities laid the framework for developing maps showing the vulnerability of ground water to contamination by the widely used herbicide atrazine (see, for example, p. 72). These maps are being used by the Idaho State Department of Agriculture to develop its State Pesticide Management Plan.

Level of needed protection increases with increasing amounts of agricultural and urban land

Concentrations of nutrients and pesticides in streams and shallow ground water generally increase with increasing amounts of agricultural and urban land in a watershed. This relation results because chemical use increases and less water is available from undeveloped lands to dilute the chemicals originating from agricultural and urban lands. In Willamette Basin streams during high spring streamflow following fertilizer application, concentrations of nitrate increased proportionately (from less than 1 up to 10 milligrams per liter) with increasing drainage area in agriculture (from about 0 to nearly 100 percent). Concentrations of nutrients also were found to increase with the percentage of drainage areas in agriculture for watersheds in the Ozark Plateaus, Potomac River Basin, and Trinity River Basin. This relation is evident not only within small watersheds but also regionally where agricultural land and chemical use extend over broad areas. For example, intensive herbicide and fertilizer use in the Upper Midwest have resulted in some of the highest concentrations of atrazine collected in stream samples across the Nation. Management strategies that are successful in reducing use and transport of this herbicide could lead to regional improvements in water quality.

The Washington State Department of Ecology recently has created a Ground Water Management Area (GWMA) to protect ground water from nitrate contamination. The GWMA covers Grant, Franklin, and Adams Counties, located in an intensive agricultural region of the Central Columbia Plateau.

Shallow ground water used for domestic supply near agricultural settings requires special consideration

Shallow ground water (less than 100 feet below land surface) in or adjacent to agricultural land use requires special consideration, particularly in rural areas where it may be used for domestic supply. The proximity to land surface and the level of human activity increase the vulnerability of this resource to contamination. Homeowners usually are not aware of potential risks because domestic wells are not monitored regularly, as is required by the Safe Drinking Water Act for large public-supply wells. In addition, many homeowners in recently established residential areas that rely on domestic wells for drinking water are not aware that chemicals leached from previously farmed land can remain in the shallow ground water for decades as a result of its slow movement.

David F. Usher

Level of protection needed in major aquifers varies with vulnerability to contamination

Varied geologic settings result in differences in vulnerability to contamination in deep major aquifers. Recognition of this can help tailor and target the appropriate level of protection and monitoring to the major aquifers of most concern, as required in the Safe Drinking Water Act source-water and drinking-water programs and in nutrient- and pesticide-management plans.

The most extensive environmental control strategies and monitoring should be considered in major aquifers in vulnerable geologic settings that allow rapid influx and vertical mixing of water from shallow ground-water systems. Such systems include sand and gravel aquifers or alluvial fans, particularly those that are heavily pumped for irrigation and public water supply, as well as karst settings that provide open conduits for relatively rapid downward movement of water. Equally important to consider are possible connections to deep parts of the aquifer through poorly constructed or improperly sealed wells that allow surface water to travel quickly down the outside of well casings.

In general, extensive environmental control strategies and monitoring are less critical for most deep aquifers when compared to shallow ground water in similar land-use settings. Water in major aquifers generally is buried and protected deep beneath the land surface. Frequent sampling is not needed because the quality of deep ground water in these aquifers is minimally affected

Bruce D. Lindsey

Concentrations of nitrate in major aquifers in the Lower Susquehanna River Basin are highest in agricultural areas in karst settings. In almost one-half of the samples, concentrations exceeded 10 milligrams per liter, the drinking-water standard for nitrate.

by seasonal events. Spatially intensive sampling generally is not needed because variations in water quality over short distances are small. Water in deep aquifers generally flows along deep and long paths that integrate water quality over large areas for extended periods, sometimes for centuries.

Even in relatively protected settings, major aquifers require some level of consideration to support long-term prevention from contamination. Ground water at all depths is part of an integrated system and can never fully escape future contamination as water moves downward from shallow systems. Future contamination in deep major aquifers could pose serious concerns because these aquifers commonly are used for public water supply and because restoration of the quality of this relatively inaccessible and slow-moving water would be costly and difficult.

Concentrations of nitrate in water from major aquifers in the Rio Grande Valley were less than 2 milligrams per liter, indicating that movement of shallow ground water into the deeper parts of the aquifer is minimal.

Streams are more vulnerable to contamination than shallow ground water in areas that are extensively tile drained and ditched

Tile drains and ditches "short circuit" the ground-water system by intercepting soil water and shallow ground water and rapidly transporting it to streams. Tiling and ditching are commonly used to drain clayey glacial sediment in parts of the Midwest and organic, clayey Coastal Plain sediment in the Southeast. Streams in these areas can have elevated concentrations of agricultural chemicals because of outflow from drains and ditches. Seepage into the ground is minimized, resulting in lower concentrations of chemicals in ground water. An awareness of these conditions can help tailor the appropriate level of management and protection to streams in these areas.

Small streams are more vulnerable to rapid and intense contamination than are larger rivers

Hydrologic and basin characteristics, including size of the basin and amount of streamflow, affect the timing of and magnitude of exposure to contaminants in the environment. Small streams respond quickly to rainfall or irrigation and, therefore, pulses of contaminants reach higher concentrations and rise and fall more quickly than in larger rivers. In contrast, larger rivers generally have more moderate levels of contaminants, but for longer durations. Recognition of these differences can help target the appropriate timing and degree of management and protection for different types of streams.

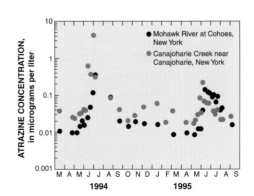

Concentrations of atrazine were 10 times higher and increased more rapidly in Canajoharie Creek than in the Mohawk River following Summer 1994 storms in the Hudson River Basin. The Mohawk River receives water not only from Canajoharie Creek but also from other tributaries draining a mix of land uses.

Development of environmental policies must consider the entire hydrologic system and its complexities, including surface-water/ground-water interactions and atmospheric contributions.

Effects of contaminants on the aquatic environment depend on surface-water flow

Contaminants and their potential effects on the environment vary throughout the year and largely depend on the amount of water flowing in a stream. Frequent monitoring is needed to characterize variations in contaminants, such as those that occur between low and high flows. Measurements of streamflow during these different conditions, in combination with water-quality samples, are needed to fully assess the amounts of contaminants transported by a stream throughout the year to a receiving body, such as an estuary. This information, particularly over the long term, is critical for developing TMDLs for streams and for assessing the potential effects of contaminants on the health and aquatic life of receiving waters.

Ground water can be a major nonpoint contributor of nutrients and pesticides to streams

Historically, ground water has been overlooked as a major nonpoint contributor of contaminants to streams and coastal waters. Ground-water issues for water managers, however, continue to grow in importance in many parts of the Nation. For example, more than one-half of the water and nutrients that enter Chesapeake Bay first travel through the ground-water system.[3] Consideration of ground-water contributions is needed in water-resource programs, such as State programs designed to establish TMDLs in streams. Exclusion of ground water may prevent a full accounting of all available sources and may limit the effectiveness that TMDLs could have in future stream restoration and protection. Consideration of ground water also may be needed to ultimately reach Clean Water Act goals for fishable, swimmable, and drinkable waters.

The significance of ground water varies with local differences in geology and soils. Ground-water contributions to streams are most significant in geologic settings that allow rapid exchange between ground- and surface-water systems. Areas underlain by karst or by permeable and well-drained sediment can undergo relatively rapid, even seasonal, exchanges of water and contaminants. As seen in agricultural areas of the Platte River Valley in Central Nebraska, high concentrations of contaminants in streams commonly seep into shallow ground water following spring applications when river flows are high. In contrast, contaminants in aquifers can flow into adjoining streams during periods of low streamflow, such as noted in the Suwannee River in Florida.

W.H. Mullins ©

Measurements of streamflow, in combination with water-quality samples, are needed to fully assess the amount of material transported by a stream to receiving bodies, such as estuarine or coastal waters.

In some areas, concentrations of contaminants may decrease as water is exchanged between streams and aquifers. For example, nitrate concentrations in about one-half of wells sampled near the South Platte River in Colorado exceeded the drinking-water standard. Ground water contributes a substantial amount of flow to the river in this area, but concentrations in the river were much lower than in the ground water because bacteria removed the nitrate as the ground water passed through the organic-rich streambed sediment.

During low-flow conditions, when inflow to a 33-mile reach of the Suwannee River, Florida, is entirely from springs and other ground water, the daily load of nitrate transported in this reach nearly doubled.

Atmospheric contributions can be significant, too

The atmosphere can be a major source of nitrogen and pesticides. More than 3 million tons of nitrogen are deposited in the United States each year from the atmosphere, derived either naturally from chemical reactions or from the combustion of fossil fuels, such as coal and gasoline. The highest contributions of nitrogen from the atmosphere occur in a broad band from the Upper Midwest through the Northeast. Recent studies have shown that as much as 25 percent of the nitrogen entering Chesapeake Bay comes from the atmosphere.[4]

Nearly every pesticide that has been investigated has been detected in air, rain, snow, or fog throughout the country at different times of year. Annual average concentrations in air and rain are generally low, although elevated concentrations can occur during periods of high use, usually in spring and summer months. Several instances have been recorded in which concentrations in rain have exceeded drinking-water standards for atrazine, alachlor, and 2,4-D.[5] Atmospheric contributions are most likely to affect stream quality during periods when direct precipitation and surface runoff are the major sources of streamflow.

The atmosphere is an important part of the hydrologic cycle that can transport nutrients and pesticides from their point of application and deposit them outside the area or basin of interest. Consideration of atmospheric contributions is critical for effective management of water resources. Because atmospheric transport can cross State boundaries, full implementation of watershed-management strategies may require State and (or) regional involvement.

Water-quality standards, guidelines, and monitoring programs should reflect environmental conditions, including seasonal variations and contaminant mixtures.

Pesticide breakdown products and contaminant mixtures present new challenges for understanding health and environmental effects of pesticides

Pesticides break down to other compounds over time in the natural environment. Little is known about the occurrence of breakdown products, or their possible health and environmental effects. Frequent detections of some breakdown products, however, indicate the need for their consideration in the development of water-quality standards and monitoring strategies. For example, the herbicide atrazine commonly breaks down to DEA (deethylatrazine) and other products, both in streams and ground water; atrazine and DEA were detected together in more than 25 percent of ground-water samples in the first 20 Study Units. Water samples without detectable parent compounds, seemingly indicating no contamination, may merely reflect chemical transformations to other compounds. In fact, studies have shown that the parent compounds metolachlor, dacthal, alachlor, and cyanazine are often less commonly found in ground-water samples than their breakdown products.[6,34]

Mixtures of contaminants also require special consideration in assessing possible health and environmental effects, and thus in developing and improving water-quality standards. More than one-half of all stream samples contained five or more pesticides, and nearly one-quarter of ground-water samples contained two or more. These mixtures of pesticide parent compounds also occur with breakdown products and other contaminants, such as nitrate. Continued research is needed to help reduce the current uncertainty in estimating risks from commonly occurring mixtures. As improved information is accumulated, the occurrence of contaminant mixtures should be considered when developing water-quality standards and monitoring requirements.

Some widely detected pesticides are not recognized in drinking-water monitoring requirements

New pesticides are introduced each year. It is often difficult to predict their behavior in the environment from laboratory experiments and to establish the appropriate level of monitoring needed to measure their occurrence. Designing appropriate monitoring programs for pesticides will, therefore, continue to be a dynamic process, continually evolving as new information is collected.

As an example, several pesticides that currently are not recognized on the USEPA Contaminant Candidate List were detected frequently in the first 20

Karen Riva-Murray

The New York State Department of Environmental Conservation is applying NAWQA pesticide information and monitoring protocols in its statewide pesticide monitoring. The NAWQA data represent a broader array of analyses and lower detection limits than data previously available. The collaborative effort was sparked by public concerns over pesticides in New York State waters and their possible relation to the incidence of breast cancer.

Study Units. The USEPA is working with the USGS to target several of these pesticides for occurrence monitoring and guidance, including health advisories, as required by the Safe Drinking Water Act. Pesticides that were commonly detected in NAWQA analyses in the first 20 Study Units but that are not currently on the contaminant list are the herbicides 2,4-D and tebuthiuron and the insecticides carbaryl, malathion, and chlorpyrifos. Although not as frequently detected in the first 20 Study Units, the herbicide acetochlor, a probable human carcinogen approved for use in 1994, also is not on the list.

Seasonal patterns dictate the timing of high concentrations in drinking-water supplies and aquatic habitats

The vulnerability to contamination of streams and ground water can differ seasonally. Increased monitoring and special management of water-supply sources may be needed during high-flow conditions and periods of agricultural chemical applications. The temporary use of ground-water sources of supply—if they have been developed and are available—might be considered as an alternative to surface-water sources to decrease the potential for not meeting drinking-water standards or aquatic-life criteria during such periods.

David F. Usher

Concentrations of nutrients and some pesticides in streams draining agricultural areas commonly are higher during spring and summer months than during the rest of the year. Chemicals generally are transported shortly after application during high-flow conditions that result from spring rains, snowmelt, and (or) irrigation. Heavy irrigation runoff, which commonly carries high concentrations of nutrients and pesticides, is of special concern in the western part of the Nation (such as in the Trinity River Basin, San Joaquin-Tulare Basins, Rio Grande Valley, and Central Columbia Plateau) because such runoff can account for the majority of streamflow.

In other parts of the Nation, patterns can be different. For example, concentrations of diazinon in streams in the San Joaquin-Tulare Basins are high during winter because of high rainfall and use of dormant sprays on orchards. Differences in patterns also result from local water-management practices, including the timing of reservoir storage and water use. Seasonal patterns must be characterized and understood for each watershed because they dictate the timing of the highest concentrations in drinking-water supplies and aquatic habitats.

Surface runoff in agricultural areas can carry eroded sediment and attached chemicals, such as DDT, to streams during periods of heavy irrigation and (or) precipitation.

Monitoring during storms is needed to track peak inputs of contaminants to streams

Excessive amounts of contaminants can enter streams during storms and can have overriding effects on the quality of streams and the respective receiving bodies, such as estuarine or coastal waters. High flows in the Susquehanna, Potomac, and James Rivers during January 1996, for example, carried nearly one-half of the phosphorus and one-quarter of the nitrogen that typically is transported to the Chesapeake Bay in an average year.[7] Fortunately, this flood occurred in winter, a time when grasses and many living organisms were dormant and when farmland, rich in nutrients, was frozen. Effects, such as increased algal growth and low levels of dissolved oxygen from subsequent algal decay, could have been much greater if the flood had occurred in spring or summer. Without monitoring information during major hydrologic events, a full accounting of nutrients and pesticides transported by streams is incomplete, and a full understanding of the effects of these contaminants on the health and living resources of receiving waters, such as the Chesapeake Bay, is restricted.

James M. Gerhart

Major events affecting streams used for drinking-water purposes may require intensified monitoring during peak fertilizer- and pesticide-application periods. As an example, the Potomac River at Washington, D.C., carried an estimated 3,300 pounds of the herbicide atrazine and 3.3 million pounds of nitrogen in 5 days during a flood in June 1996. On two consecutive days following the storm, atrazine was measured at concentrations greater than the drinking-water standard of 3 micrograms per liter.

Considerations in monitoring the effectiveness of conservation buffers

Conservation buffers are small areas or strips of vegetation designed to mitigate the movement of sediment, nutrients, and pesticides within and from farm fields. They are supported by the U.S. Department of Agriculture (USDA) Farm Bill and many conservation programs, such as the Conservation Reserve Program, Environmental Quality Incentives Program, and the Stewardship Incentives Program. The USDA goal is to help landowners install 2 million miles of conservation buffers by the year 2002.[8]

There are two considerations in monitoring the effectiveness of conservation buffers. The first consideration relates to tracking ground-water quality. In some areas, slowing the transport of runoff to streams by use of conservation buffers can increase infiltration of water and contaminants into the ground. As shown by the USGS in the Delmarva Peninsula, a pilot NAWQA study initiated in the mid-1980s, the transport and fate of these contaminants in the ground is variable, depending on soil and aquifer composition, topography, and rates and pathways of ground-water flow.[9] Monitoring of ground-water quality might, therefore, be beneficial to fully assess potential effects of conservation strategies.

The second consideration relates to time of year and its implications on tracking stream quality. The effectiveness of conservation buffers on stream quality is likely to be most evident when streamflow is dominated by runoff from rainfall, snowmelt, and (or) irrigation following chemical applications. Their effectiveness is likely to be less evident during low-flow conditions, when most of the streamflow is from ground-water discharge.

Vegetation along waterways can help slow surface runoff and movement of nutrients, pesticides, and sediment within and from farm fields and can improve stream quality.

Long-term monitoring may be needed to evaluate the effectiveness of crop-management practices

Long-term monitoring may be needed to evaluate the effectiveness of some environmental control strategies, such as crop-management practices, because of the slow rate of ground-water flow and the time lag between adoption of practices and improvement of water quality. As demonstrated in the San Joaquin-Tulare Basins, shallow ground water below farmland will improve first, sometimes in several years or less. Decades may pass, however, before water quality improves in deeper aquifers.

A time lag between adoption of crop-management practices and improvement of water quality also can occur for streams. Because ground water containing elevated concentrations of nutrients and pesticides can discharge to surface water, enhancement of stream quality also could lag changes in agricultural practices by years or decades.

National findings and their implications for water policies and strategies

Consistent and systematic information is needed over the long term to measure local, regional, and national trends

For many chemicals, it is too early to assess trends because historical data are insufficient or inconsistent. Some trends have emerged, however, from monitoring nutrients and pesticides; they show that changes in water quality over time are controlled largely by soils, geology, and other natural features, and by changes in chemical use and management practices. For example, concentrations of phosphorus and ammonia have decreased in rivers downstream from wastewater treatment plants since the 1970s because of improved treatment technology. Concentrations of organochlorine insecticides have been reduced in sediment and fish since restrictions on production and distribution of these pesticides in the 1970s and 1980s.

Changes in concentrations of modern, short-lived pesticides follow changes in use and tend to be focused in specific regions and land-use areas. For example, increases in acetochlor and decreases in alachlor are evident in some streams in the Upper Midwest, where acetochlor began replacing alachlor for control of weeds in corn and soybeans in 1994. The changes in use are reflected in stream quality relatively quickly, generally within 1 to 2 years.

In contrast, ground-water quality responds more slowly to changes in chemical use or adoption of land-management practices, typically lagging by several years and even decades. Local variations in natural features, such as soil types and amounts of recharge, can result in variable rates of ground-water flow, which thereby affect long-term responses to land-management practices. For example, concentrations of nitrate decreased significantly (from about 18 milligrams per liter in the mid-1980s to less than 2 milligrams per liter in the mid-1990s) in ground water underlying parts of the Central Platte Natural Resources District, Nebraska, after implementation of fertilizer management strategies. Yet, despite implementation of the strategies, the response has been delayed in other parts of the District because of differences in local features controlling ground-water flow. Specifically, concentrations of nitrate remained greater than two times the USEPA drinking-water standard in nearly one-fourth of wells in one area sampled by the District in the mid-1990s.

Systematic and consistent monitoring over the long term is essential at local, State, and national levels. Such monitoring will help water managers and policy makers to evaluate how well local and regional environmental controls are working and to choose the most cost-effective resource strategies for the future.

David F. Usher

Progress in water-quality improvement, especially in ground water, may not be evident for years after farmers change their land-management practices because of slow ground-water movement.

Reliable predictive models are required to cost effectively estimate water-quality conditions that can not be directly measured for a wide range of possible circumstances.

Effective strategies for managing nutrients and pesticides, as well as related water-quality issues, require far more information than we can afford to directly measure for the full range of places and times that are important. Moreover, many management problems, ranging from deciding how much to spend on a management strategy to approving a pesticide for use, are inherently related to predicting potential effects on water quality. Models and other methods can be useful for predicting water-quality conditions over a wide range of possible circumstances and are essential for improving water-quality management over broad regions.

NAWQA findings are beginning to play an important role in model development and validation, and an increased emphasis of explanatory and predictive modeling is planned for the second cycle of investigation in each Study Unit. Early examples are the estimation of ground-water vulnerability to atrazine contamination in the Upper Snake River Basin (p. 72) and to nitrate contamination in the Puget Sound Basin (see sidebar). In addition, ground-water vulnerability to nitrate contamination also was assessed at the national scale (p. 51). Although not directly predicting an outcome, these analyses use correlations to rank the likelihood and risk of contamination.

One of the most important roles that NAWQA can fulfill in working with water-management agencies is to provide systematic, high-quality data that can be used to develop and test predictive models for hydrologic systems throughout the Nation. The USEPA, for example, is using NAWQA pesticide data to test the reliability of models now being used to predict possible pesticide occurrence in streams and reservoirs. Water-quality models have been in use for many years, but their utility depends on their reliability for representing actual conditions. Without demonstrated reliability based on comparisons to measured conditions, confidence in a model is difficult to attain, and the usefulness of the model in decision making, especially in controversial situations, is limited.

As NAWQA studies progress from an emphasis on assessing and documenting water-quality conditions and cause-and-effect factors (during the first cycle of investigation) to an emphasis on a more detailed understanding of the most critical processes controlling water quality (during the second cycle), the development of predictive models will continue to grow and play a more vital role in both analysis and water-management applications.

Predicting ground-water vulnerability to nitrate contamination

A statistical model was created to predict the vulnerability of ground water to nitrate contamination from human activities in urban and agricultural areas in the Puget Sound Basin, Washington.[10] Factors that were used to predict the risk of contamination were well depth, surficial geology, and the percentage of agricultural and urban land use within a 2-mile radius of the well. Results from risk models provide managers with tools for guiding future land-use development, assessing potential health risks associated with nitrate, and designing cost-effective monitoring programs.

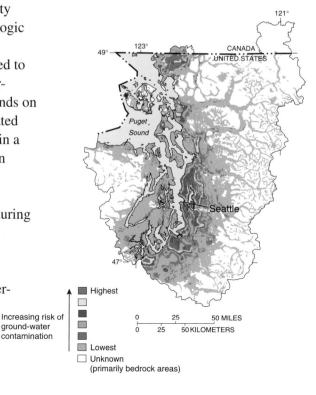

Increasing risk of ground-water contamination

Highest

Lowest

Unknown (primarily bedrock areas)

Point sources are regulated

Point sources are regulated by laws that place limits on the types and amounts of contaminants released to water. Legislation has resulted in reductions in industrial sources and upgrades to wastewater treatment plants. Although violations occur, the legislation has had a positive influence on preventing or limiting contaminants from entering water systems.

Nonpoint sources contribute more contaminants than point sources

It is more difficult to develop solutions for nonpoint sources, which are vastly more widespread and difficult to identify and quantify than point sources. For example, in the first 20 Study Units, it is estimated that about 90 percent of nitrogen and 75 percent of phosphorus originates from nonpoint sources; the remaining percentages are from point sources.

The atmosphere is a nonpoint source of contamination

The atmosphere commonly is overlooked as a source of nutrient and pesticide contamination. Yet, more than 3 million tons of nitrogen are deposited in the United States each year from the atmosphere. The nitrogen is derived either naturally from chemical reactions or from the combustion of fossil fuels, such as coal and gasoline. Local contributions also come from evaporative losses of nutrients in the vicinity of open-air manure lagoons. Nearly every pesticide that has been investigated has been detected in air, rain, snow, or fog across the Nation at different times of year. Atmospheric deposition is not evenly distributed across the United States. For example, the highest deposition rates of nitrogen (greater than 2 tons per square mile) occur in a broad band from the Upper Midwest through the Northeast.

Nonpoint-source contamination comes from many diffuse sources, including fertilizers and pesticides from agricultural and residential lands, and nutrients from livestock and pet waste and from septic systems. Nonpoint-source contamination starts with precipitation falling on the ground. As the resulting runoff moves over and through the soil, it picks up and carries away natural and human-made contaminants, finally transporting them to streams, rivers, wetlands, lakes, and even ground water. Nonpoint-source contamination is the leading and most widespread cause of water-quality degradation, and it can have harmful effects on drinking-water supplies, recreation, fisheries, and wildlife.

Streams, which commonly serve as drinking-water supplies, recreational areas, and biological habitats, are influenced by the landscape through which they flow and the atmosphere from which the water originates. The vulnerability of streams to contamination reflects a complex combination of upstream natural processes, land use, chemical use, and land-management practices.

Assessment of the occurrence of nutrients and pesticides in water resources requires recognition of complicated interconnections among surface water and ground water, atmospheric contributions, natural landscape features, human activities, and aquatic health. The vulnerability of surface water and ground water to degradation depends on a combination of natural landscape features, such as geology, topography, and soils; climate and atmospheric contributions; and human activities related to different land uses and land-management practices.

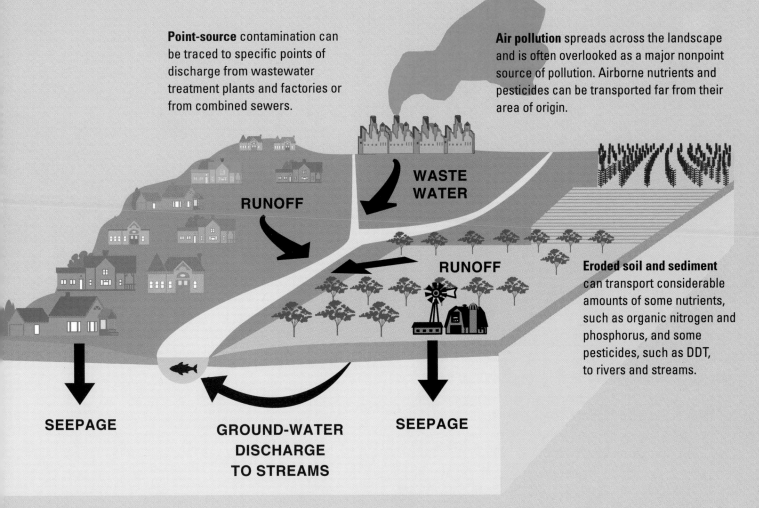

Point-source contamination can be traced to specific points of discharge from wastewater treatment plants and factories or from combined sewers.

Air pollution spreads across the landscape and is often overlooked as a major nonpoint source of pollution. Airborne nutrients and pesticides can be transported far from their area of origin.

WASTE WATER

RUNOFF

RUNOFF

Eroded soil and sediment can transport considerable amounts of some nutrients, such as organic nitrogen and phosphorus, and some pesticides, such as DDT, to rivers and streams.

SEEPAGE

GROUND-WATER DISCHARGE TO STREAMS

SEEPAGE

Fish and other aquatic organisms reflect cumulative effects of water chemistry and land-use activities. Fish, for example, acquire some pesticides by ingesting stream invertebrates or smaller fish that have fed on contaminated plants. Fish also can accumulate some contaminants directly from water passing over their gills.

Ground water—the unseen resource—is the source of drinking water for more than 50 percent of the Nation. As water seeps through the soil, it carries with it substances applied to the land, such as fertilizers and pesticides. Water moves through water-bearing formations, known as aquifers, and eventually surfaces in discharge areas, such as streams, lakes, and estuaries. It is common to think of surface water and ground water as separate resources; however, they are interconnected. Ground-water discharge can significantly affect the quality and quantity of streams, especially during low-flow conditions. Likewise, surface water can affect the quality and quantity of ground water.

Sources of nutrients and pesticides

About 1 billion pounds of pesticides are applied each year

Total pesticide use in the United States has remained relatively constant at about 1 billion pounds per year, after growing steadily through the mid-1970s because of increased use of herbicides. Agriculture now accounts for 70 to 80 percent of total pesticide use. Most agricultural pesticides are herbicides, which account for about 60 percent of the agricultural use. Insecticides generally are applied more selectively and at lower rates than herbicides. Major changes in insecticide use have occurred over the years in response to environmental concerns, which have resulted in various restrictions on the use of organochlorine insecticides, such as DDT. Specifically, as the use of these persistent pesticides declined, the use of other, less persistent insecticides increased.

Commercial fertilizer and manure are important sources of nutrients

About 12 million tons of nitrogen and 2 million tons of phosphorus are applied each year as commercial fertilizer. Another 7 million tons of nitrogen and 2 million tons of phosphorus are applied as manure. The distribution of fertilizer use varies across the Nation. The highest application rates (greater than 7 tons per square mile) occur over a broad area of the Upper Midwest; other areas of high application are along the East Coast, throughout the Southeast, and in agricultural areas of the West. Private septic systems also can be important sources of nutrients in rural and residential areas.

Phosphorus contributions from laundry detergent have decreased

From about 1940 to 1970, laundry detergent was a major source of phosphorus to the environment. Contributions decreased to almost negligible amounts after the enactment of State phosphate detergent bans beginning in the 1970s and the voluntary cessation of phosphate use by detergent manufacturers.[11]

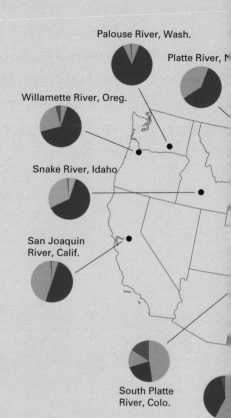

Palouse River, Wash.

Platte River, N

Willamette River, Oreg.

Snake River, Idaho

San Joaquin River, Calif.

South Platte River, Colo.

Rio Gr
N. Me

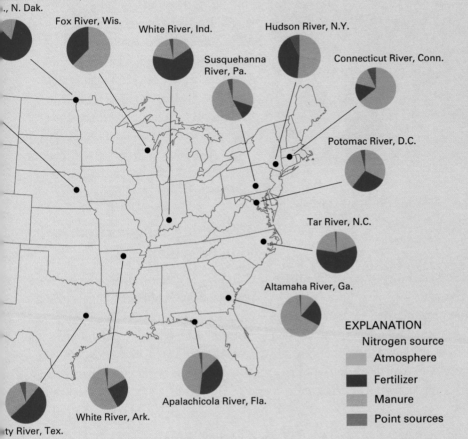

River of the North,
..., N. Dak.

Fox River, Wis.

White River, Ind.

Hudson River, N.Y.

Susquehanna
River, Pa.

Connecticut River, Conn.

Potomac River, D.C.

Tar River, N.C.

Altamaha River, Ga.

Apalachicola River, Fla.

White River, Ark.

ty River, Tex.

EXPLANATION

Nitrogen source

Atmosphere

Fertilizer

Manure

Point sources

Proportions of nonpoint and point sources of nitrogen vary in watersheds across the continental United States.[12] Commercial fertilizer and manure typically constitute the major sources of nitrogen to the first 20 NAWQA Study Units. Atmospheric nitrogen is significant in most Study Units except in the far West and the Northern Great Plains. Point sources are an important source of nutrients to watersheds near large urban areas, such as Denver in the South Platte River Basin and Hartford in the Connecticut, Housatonic, and Thames River Basins.

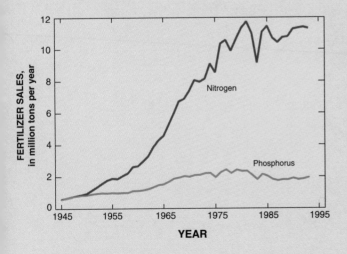

The use of nitrogen and phosphorus fertilizer increased twentyfold and fourfold, respectively, between 1945 and the early 1980s, and then leveled off through 1993.

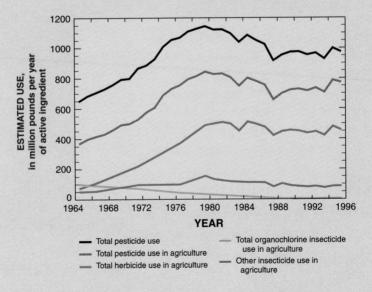

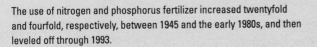

Total pesticide use

Total pesticide use in agriculture

Total herbicide use in agriculture

Total organochlorine insecticide use in agriculture

Other insecticide use in agriculture

Changes in agricultural pesticide use over the decades.[13–17]

Focusing the NAWQA study design on land use

The NAWQA study design focuses on streams and shallow ground water in specific land-use settings to evaluate individual types of nonpoint sources of nutrients and pesticides, and on major rivers and aquifers to evaluate the effects of these sources at larger scales.[5] Major rivers and aquifers usually integrate water-quality effects from complex combinations of land uses and environmental settings within large contributing areas.

Targeted land uses include agricultural, urban, and, to a lesser degree, undeveloped forest or rangeland settings. Most of the agricultural streams studied had watersheds with more than 70 percent cropland and pasture, and all wells sampled in studies of shallow ground water in agricultural areas were within targeted cropland or pasture areas. Studies of streams and shallow ground water in urban areas focused primarily on residential land with low to medium population densities (300 to 5,600 people per square mile).[2] Many of the larger river sites in mixed land-use settings, however, were down-stream from major metropolitan areas with important point sources, such as discharges from municipal and industrial wastewater treatment plants.

Collecting samples from streams

Water samples were collected from a total of 212 stream sites for nutrient analysis and a subset of 65 of these sites for pesticide analysis.[18] Samples of bed sediment were collected at 521

Yakima NAWQA Staff

sites for analysis of organochlorine insecticides.[19] Fish were sampled for analysis of the same group of compounds at about half of the sites where bed sediment was sampled (see table below).[20]

Water samples were collected from streams throughout the year, including high-flow and low-flow conditions. Samples were collected for nutrient analysis approximately once each

month, and more frequently during high-flow periods. At sites where pesticide samples were collected, sampling was more intensive for both nutrients and pesticides—generally weekly or twice monthly for a 4- to 9-month period during the time of highest chemical use and runoff. For most sites sampled for bed sediment or fish, a single sample was collected during low-flow conditions.

Collecting samples from ground water

Samples of shallow ground water were collected for analysis of nutrients and pesticides as part of ground-water studies, each of which commonly consisted of 20 to 30 randomly selected wells within the targeted land use.[21, 22] Most wells were completed

Summary of stream sampling sites and ground-water studies

	Nutrients	Pesticides
Relatively Undeveloped Forest or Rangeland		
Streams	28	—
Bed sediment	—	83
Shallow ground-water studies	4	some wells
Agricultural Land		
Streams	75	40
Bed sediment	—	173
Shallow ground-water studies	36	36
Urban Land		
Streams	22	11
Bed sediment	—	71
Shallow ground-water studies	13	13
Mixed Land Use		
Streams and rivers	87	14
Bed sediment	—	194
Major aquifers	33	33

just below the water table, and most were less than 100 feet deep. Ground water in these shallow zones is more susceptible to degradation from human sources and activities than deeper ground water.

For major aquifers, samples were collected primarily from existing private domestic wells. Similar to studies of shallow ground water, 20 to 30 wells were randomly selected in

Delmarva NAWQA Staff

most aquifers, but without regard to land use. Most wells were sampled one time, and data analyses are based on one sample per well.

Mapping national distributions of water quality

Nutrient and pesticide levels are summarized on U.S. maps to facilitate analysis and comparison of regional and national patterns. Concentrations or detection frequencies are ranked according to three categories: *lowest* for the lowest 25 percent, *medium* for the middle 50 percent, and *highest* for the highest 25 percent of concentrations or detection frequencies among all stream sites or ground-water studies.

For nutrients in streams, flow-weighted total nitrogen and total

phosphorus concentrations were determined for each stream site for 1994 and 1995 and were averaged.[23] For shallow ground water, median nitrate concentration was determined for each ground-water study.

For pesticides in water at each stream site, concentrations were summed separately for all detected herbicides and for all detected insecticides in each water sample. Then, annual results (75th percentile of monthly median concentrations) for total herbicides and total insecticides were determined for each site as the basis for ranking. For organochlorine insecticides, the distribution in bed sediment was mapped instead of the distribution in fish because data are available for a larger number of bed sediment sites. Concentrations of all organochlorine insecticides were summed for each site as the basis for ranking. For pesticides in ground water, each ground-water study area was ranked on the basis of the frequency of detection of one or more herbicides, or one or more insecticides, in the wells sampled.

The maps of national results for nutrients and pesticides also show patterns of nonpoint inputs of nitrogen, phosphorus, and pesticides. Based on county agriculture statistics for 1987 and 1992, average annual nitrogen and phosphorus inputs to agricultural and urban land were estimated from commercial fertilizer sales (1991–94) and manure from animals (1992). Average annual input of nitrogen from the atmosphere was estimated from 1991–94 data.[24] Use estimates for currently used herbicides

and insecticides were made for every county in the Nation on the basis of crops harvested in 1992 and typical pesticide use rates in the early 1990s (http://water.usgs.gov/lookup/get?nawqapest/use92/mapex.html). The estimated use was then mapped for agricultural land throughout the Nation. The same general method was used for organochlorine insecticides, except that crop data from 1978 and use rates from 1966 were used.

Assessing potential effects of contaminants on human health and the environment

In this report, potential effects of nutrients and pesticides have been assessed by comparing concentrations to available standards and guidelines for protection of human health, aquatic life, or wildlife.[25] "Standards" are legally enforceable, whereas "guidelines" are primarily advisory. Standards and guidelines provide useful and widely used benchmarks that serve as starting points for evaluating potential effects of contaminants in the environment. Stream sites are flagged if one or more water-quality guidelines for protection of aquatic life were exceeded in one or more samples. Ground-water study areas are flagged if one or more human-health standards or guidelines were exceeded in one or more wells. More details about how specific nutrient and pesticide concentrations are compared to standards and guidelines are described in the nutrients and pesticides sections.

Dennis A. W

Nutrients from atmospheric and urban sources, fertilization, and livestock wastes can contribute to excessive algal growth in streams

Anonym

Kevin D. Richa

Jeffrey D. Martin

Ian R. Wa

Nutrients

Human activities—including agricultural and urban uses of fertilizer, agricultural use of manure, and combustion of fossil fuels—have caused widespread increases of nitrate in shallow ground water and total nitrogen and total phosphorus in streams across the Nation.

Nitrate did not pose a national health risk for residents whose drinking water came from streams or from major aquifers buried relatively deep beneath the land surface. Some concerns were evident in 4 of the 33 major aquifers sampled, where nitrate concentrations in more than 15 percent of each aquifer exceeded the USEPA drinking-water standard. The most prevalent nitrate contamination of ground water, however, was found in relatively shallow ground water in rural areas where the water commonly is used for domestic supply.

In more than one-half of sampled streams, concentrations of total nitrogen and total phosphorus were above national background concentrations. Elevated phosphorus levels, in particular, can lead to excessive plant growth (eutrophication) in freshwater environments; in more than one-half of sampled streams and in three-fourths of agricultural and urban streams, average annual concentrations of total phosphorus exceeded the USEPA desired goal for prevention of nuisance plant growth. The highest total nitrogen and total phosphorus concentrations were found in small streams draining watersheds with large proportions of agricultural or urban land. Long-term monitoring of streams indicates that programs to control point-source discharges of phosphorus and ammonia have been effective, despite population increases in most metropolitan areas. Phosphorus concentrations have decreased as a result of reductions in the use of phosphate detergents and in the amount of phosphorus discharged from upgraded wastewater treatment plants. Improved wastewater treatment, which converts ammonia to nitrate, generally has resulted in a decrease in ammonia concentrations and an increase in nitrate concentrations in streams. Thus, concentrations of total nitrogen downstream from metropolitan areas have changed little during the past 20 years, although toxicity to fish has decreased with decreasing ammonia levels.

Results from NAWQA studies have shown regional and seasonal differences in nutrient concentrations that can be explained largely by the amounts and timing of fertilizer and manure applications and by the variety of soils, geology, climate, and land- and water-management practices across the Nation. Recognition of these differences is important for efficient protection of ground water needed for drinking and for curbing eutrophication of surface water. ◗

WHAT WAS MEASURED...

Nitrate is the primary form of nitrogen dissolved in streams and ground water. In this report, nitrate refers to the sum of nitrate plus nitrite, as reported by the USGS laboratory. Nitrite concentrations commonly were less than the laboratory detection level of 0.01 milligrams per liter (mg/L), making its contribution to nitrate plus nitrite negligible.

Ammonia is a dissolved form of nitrogen that is less common than nitrate. As measured by the USGS laboratory, total ammonia includes ammonium ion and **un-ionized ammonia**. The latter is usually a minor component of ammonia at pHs commonly observed in streams and ground water.

Total nitrogen includes nitrate, nitrite, ammonia, and organic nitrogen. Nitrite is generally unstable in surface water and contributes little to the total nitrogen. Organic nitrogen (mostly from plant material or organic contaminants) can exist in considerable proportions and contribute substantially to total nitrogen in streams.

Phosphates are the most common forms of phosphorus found in natural waters. Compared to nitrate, phosphates dissolve less readily. They are not mobile in soil water and ground water because they tend to attach to soil and aquifer particles. They can have a significant impact, however, because eroded soil can transport considerable amounts of attached phosphates to streams and lakes. **Orthophosphate** typically constitutes the majority of dissolved phosphates, which can be readily assimilated by aquatic plants and promote eutrophication.

Total phosphorus includes phosphates, as well as all other phosphorus forms. Dissolved phosphates and particulate organic phosphorus (mostly from plant material) are the main components of total phosphorus.

For brevity, all forms of nutrients discussed in this report represent concentrations as either nitrogen or phosphorus. For example, a nitrate concentration expressed as 10 mg/L refers to a nitrate concentration of 10 mg/L as nitrogen.

Background concentrations of nutrients are low in streams and ground water

Background concentrations of nutrients were estimated on the basis of samples collected from undeveloped areas considered to be minimally affected by agriculture, urbanization, and associated land uses. Background concentrations in undeveloped areas are controlled primarily by naturally occurring minerals and by biological activity in soil and streambed sediment. Chemical properties of the atmosphere and rainwater, which can reflect human-related fuel combustion and other activities both within and external to a watershed, can increase background concentrations.

In this report, national background nutrient concentrations include atmospheric contributions and are summarized in the following table. **Waters with concentrations of nutrients greater than the national background concentrations are considered to have been affected by human activities in a variety of land-use settings.**

ESTIMATES OF NATIONAL BACKGROUND NUTRIENT CONCENTRATIONS

Nutrient	Background concentration (mg/L)
Total nitrogen in streams [Data from 28 watersheds in first 20 Study Units]	1.0
Nitrate in streams [26]	0.6
Ammonia in streams [26]	0.1
Nitrate in shallow ground water [27]	2.0
Total phosphorus in streams [26]	0.1
Orthophosphate in shallow ground water [Data from 47 wells in first 20 Study Units]	0.02

Background nutrient concentrations can vary considerably from region to region, or even within watersheds, because of differences in hydrology and in naturally occurring nutrient levels in soils, rocks, and the atmosphere. The data analyzed for this report are insufficient to define background nutrient concentrations on a regional basis. Thus, all available data from undeveloped areas were combined to derive national background concentrations. The national background concentrations are higher than most concentrations measured in relatively undeveloped areas across the Nation and may not be applicable for use in regional or local analyses.

Human activities have increased nutrients above background concentrations

Effects of human activities on nutrients were assessed by comparing concentrations in streams and ground water to national background concentrations. Waters with nutrient concentrations above background are referred to as "enriched" in this report. Fifty-seven percent of sampled streams were enriched with total phosphorus on the basis of average annual total phosphorus concentrations exceeding national background concentrations. Similarly, 61 percent of sampled streams were enriched with total nitrogen and nitrate, but only 23 percent of sampled streams were enriched with ammonia. Only 1 of 28 relatively undisturbed forested or rangeland streams had average annual concentrations of total phosphorus or total nitrogen above national background concentrations. Most of the streams that were enriched with nutrients drained areas of agricultural and (or) urban land.

In 53 percent of shallow ground-water studies in agricultural and urban areas, median nitrate concentrations were above the national background concentration. Median nitrate concentrations were above background in only 3 of 33 major aquifers studied. Those three aquifers were beneath agricultural areas in three different Study Units.

In most cases, enrichment of streams with nutrients occurred in small watersheds and (or) regions dominated by agricultural or urban land use. Effects of human activities were found in shallow ground beneath agricultural and urban areas throughout the Nation, but not in many of the major aquifers sampled.

STANDARDS AND GUIDELINES FOR PROTECTING WATER QUALITY

The USEPA has established a Federal drinking-water standard or Maximum Contaminant Level (MCL) of 10 mg/L for **nitrate**.[28] An MCL is a concentration above which adverse human health effects may occur.

The USEPA[29] has established criteria for **un-ionized ammonia** in surface water because of its toxicity to fish. The chronic criteria vary from 0.07 to 2.1 mg/L of total ammonia for pHs of 6.5–9.0 and water temperatures of 0–30 °C.

National criteria have not been established for concentrations of dissolved phosphates in streams or ground water.

National criteria have not been established for total phosphorus or total nitrogen in streams. The USEPA has established a desired goal of 0.1 mg/L **total phosphorus** for the prevention of nuisance plant growth in streams and other flowing waters not discharging directly to lakes or impoundments.[29]

The Otter Tail River, Minnesota (above), which supports a healthy growth of wild rice, and Fir Creek, Oregon (right), which contributes to Portland's drinking-water supply, are examples of streams with low nutrient concentrations.

Jeffrey D. Stoner

Dennis A. Wentz

Nutrients are a potential concern for human health

From a national perspective, nitrate contamination did not pose a health risk for residents who drank water from major aquifers buried relatively deep below land surface. Some concerns were evident, however, in 4 of the 33 major aquifers sampled. In each of these four aquifers, nitrate concentrations in more than 15 percent of samples exceeded the USEPA drinking-water standard. All four aquifers are relatively shallow, in agricultural areas, and composed of sand and gravel that is vulnerable to contamination by land application of fertilizers. In nearly one-half of the major aquifers sampled, water from at least one well, out of 20 to 30 wells, exceeded the drinking-water standard. **Many of the major aquifers exhibiting high nitrate concentrations were used for rural domestic water supply.**

A national ranking of NITRATE concentrations in major aquifers

Percentage of samples exceeding drinking-water standard for nitrate (10 milligrams per liter)—
Each circle represents a major aquifer

- ● Greater than 15
- ● Less than 15 (but at least 1 sample)
- ○ 0 samples exceed standard

Background concentration

- ○ Bold outline indicates median values greater than background concentration (2 milligrams per liter)

See p. 31 for more information about this map

About 15 percent of all shallow ground water sampled beneath agricultural and urban areas exceeded the drinking-water standard for nitrate. **The presence of elevated nitrate concentrations in shallow ground water raises concerns, particularly where these vulnerable aquifers contain deeper wells used for rural domestic water supply. Contamination of shallow ground water may be a warning to alert populations to potential future risks from consumption of water from deeper wells in these aquifers.**

Because of its proximity to the land surface, shallow ground water is younger and more vulnerable to contamination from human activities than deep ground water. Major aquifers generally are buried deep beneath the land surface, where they are protected by layers of clay or rock that are relatively impermeable and that impede downward movement of water and nitrate. Ground water in major aquifers sampled by the NAWQA Program can be tens to hundreds of years old and, therefore, minimally affected by recent land-use practices.

Geology and hydrology control the movement of contaminated water from shallow to deep systems, and understanding their effects allows an anticipation of possible areas of concern in major aquifers. For example, elevated concentrations of nitrate were detected in a major aquifer in the Lower Susquehanna River Basin (median concentration about 7 mg/L) because karst (weathered carbonate rock) contains open conduits that allow rapid downward movement of water and chemicals. Median concentrations of nitrate also were high in the alluvial aquifer of the Central Nebraska Basins (about 6 mg/L) and in the alluvial fans of the San Joaquin-Tulare Basins (about 5 mg/L). Extensive pumping causes vertical mixing of ground water in these relatively permeable sand and gravel aquifers. In contrast, the median concentration of nitrate was low (0.4 mg/L) in a surficial sand and gravel aquifer in the Red River of the North Basin; however, 15 percent of the samples exceeded the drinking-water standard. This last example demonstrates the complex effects that local geology and nitrate sources can have on nitrate contamination.

For large rivers and most of the smaller streams sampled, nitrate is not a drinking-water issue. This conclusion is based on comparisons of average annual and average monthly concentrations to the drinking-water standard. In only two of all sampled rivers and streams did the average annual concentration of nitrate exceed the drinking-water standard. These two streams drained small agricultural watersheds in the Lower Susquehanna River and Willamette Basins, and neither was used to supply drinking water.

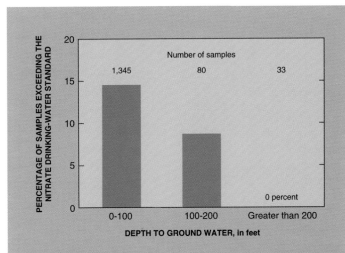

The percentage of ground-water samples with concentrations of nitrate exceeding the drinking-water standard of 10 mg/L decreases as depth to water increases. Mixing of shallow ground water with deeper, uncontaminated water and increased thickness of protective, impermeable geologic materials with depth may help explain this relation.

From a national perspective, nitrate did not pose a health risk for residents who drank water from major aquifers buried relatively deep below the land surface. People drinking ground water from shallow wells in vulnerable geologic settings (sand, gravel, or karst) in rural agricultural areas, however, are at risk of exposure to nitrate contamination.

Nutrients are a potential concern for aquatic life

Average annual and average monthly concentrations of un-ionized ammonia did not exceed USEPA aquatic-life criteria for most streams sampled. Exceptions include an agricultural stream affected by upstream wastewater treatment plant effluent in the San Joaquin-Tulare Basins and two urban streams, one in the South Platte River Basin and another in the Nevada Basin and Range. The urban streams, which are in relatively arid climates and exceeded the criteria year-round, also received effluent from wastewater treatment plants.

Eutrophic conditions were noted in some streams across the Nation because of elevated concentrations of nutrients. For example, the average annual concentration of total phosphorus in 57 percent of all streams sampled was greater than the USEPA desired goal of 0.1 mg/L for preventing nuisance plant growth in streams. In addition, about 75 percent of agricultural and urban streams exceeded this goal. **It is difficult and premature, however, to attempt a national summary of eutrophication effects because of limited available methodologies for deriving criteria based only on nutrient concentrations.** Moreover, the uncertainty regarding how nutrient contamination of streams harms aquatic life and affects nuisance plant growth does not lessen the value of accurate information for management of our Nation's streams. A strategy, spearheaded by the USEPA in collaboration with other Federal and State agencies, is underway to evaluate excessive aquatic plant growth, such as algae, in surface water. This strategy includes an understanding of stream nutrient dynamics, stream habitat (including shading and temperature), turbidity, and algal-growth processes.

Nitrogen and phosphorus have different effects on aquatic plant growth in freshwater and saltwater. Eutrophication of freshwater streams generally results from high phosphorus concentrations. In contrast, excess nitrogen, and nitrate in particular, can lead to algal blooms in coastal waters. The USEPA suggests a desired goal of 0.1 mg/L total phosphorus for freshwater streams, but there are no national criteria established for nitrogen concentrations to control excessive aquatic plant growth in coastal bays and estuaries.

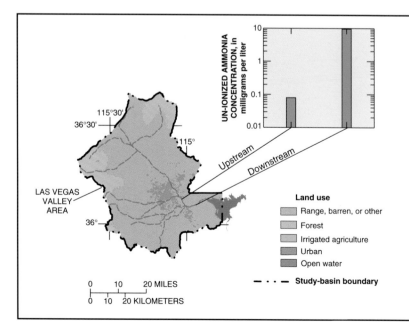

Un-ionized ammonia concentrations exceeded USEPA criteria for protection of aquatic life in Las Vegas Wash, Nevada, downstream from wastewater treatment plant discharges. Concentrations in all samples collected from April 1993 to April 1995 exceeded the criteria. The median ammonia concentration downstream from wastewater treatment plant discharges was more than 100 times the median value upstream from the discharges. Downstream ammonia concentrations decreased fivefold during 1996–97, following full implementation of tertiary treatment of wastewater, and USEPA criteria probably are no longer exceeded at this site.

"As part of the Clean Water Action Plan, the Vice President called upon USEPA to accelerate development of nutrient water-quality criteria for beneficial ecological uses in every geographic region in the country. We will work with States and tribes to develop a methodology for deriving criteria, as well as developing criteria where data are available, for nitrogen and phosphorus runoff for lakes, rivers, and estuaries by the year 2000. We intend to develop such criteria on a regional basis using scientifically defensible data and analysis of nutrients, such as those available from the USGS. We will assist States and tribes in adopting numerical nutrient criteria as water-quality standards by the end of 2003."

Robert Cantilli, Nutrients Criteria Coordinator, USEPA

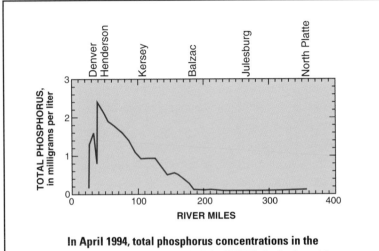

In April 1994, total phosphorus concentrations in the South Platte River exceeded the USEPA desired goal for preventing plant nuisances (0.1 mg/L) in a 150-mile reach downstream from Denver, Colorado.

Large amounts of nitrate enter the Chesapeake Bay from the Susquehanna River. Nitrate concentrations in the Susquehanna River at Harrisburg, Pennsylvania, generally were less than 2 mg/L. However, these concentrations, when multiplied by the large flows of the Susquehanna River, contribute large amounts of nitrate to Chesapeake Bay (especially compared with other rivers entering the bay) and provide enough nitrate to stimulate algal growth and affect the bay ecosystem.

Larry J. Puckett

Nutrient conditions differ by land use

The highest nitrogen and phosphorus concentrations generally were found in agricultural and urban streams. Nutrient concentrations in areas of mixed land use were lower than in agricultural or urban areas but were higher than in undeveloped areas. Orthophosphate concentrations in ground water typically were so low that relations to land use are not definitive.

Except for nitrogen in agricultural areas, nutrient concentrations in streams generally were higher than those in shallow ground water, regardless of land use. In agricultural areas, nitrate concentrations in shallow ground water typically were higher than total nitrogen concentrations in streams, although exceptions occurred in areas where soil characteristics restrict downward movement of water.

Regional patterns in nutrient concentrations can be useful for determining areas of the Nation where environmental settings deserve the greatest concern and attention. The discussion and maps on pages 41–45 focus on geographic patterns in relation to land use and nutrient inputs. Methods used to construct the maps are explained on page 31.

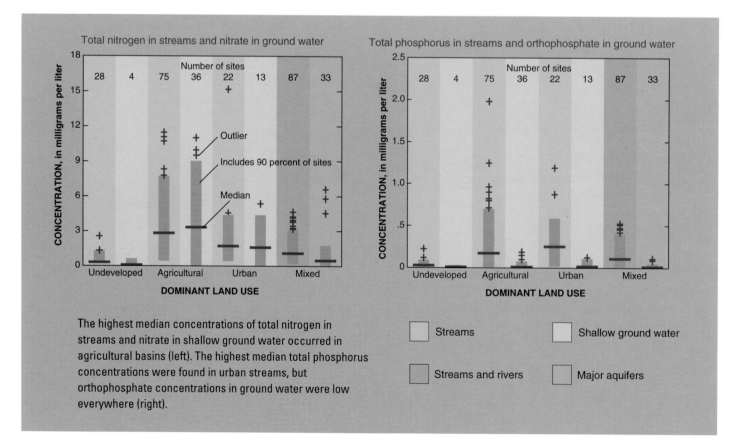

The highest median concentrations of total nitrogen in streams and nitrate in shallow ground water occurred in agricultural basins (left). The highest median total phosphorus concentrations were found in urban streams, but orthophosphate concentrations in ground water were low everywhere (right).

NITRATE IN SHALLOW GROUND WATER

High concentrations of nitrate in shallow ground water were widespread and strongly related to agricultural land use, but there were no apparent regional patterns. Based on comparisons with background concentrations, human activities have increased nitrate concentrations in ground water for about two-thirds of agricultural areas studied, compared to about one-third of urban areas. Median nitrate concentrations for 13 of 36 agricultural areas were greater than 5 mg/L and ranked among the highest of all shallow ground-water studies. Only 1 of 13 urban areas fell into this high-concentration group. It is likely that nitrogen sources in urban areas are relatively localized when compared with the generally more intensive and widespread use of fertilizers on cropland. Also, the impervious surfaces typically found in urban areas generally result in surface runoff of nutrient-laden water, rather than seepage to ground water.

A national ranking of NITRATE concentrations in shallow ground water

Median concentration of nitrate—in milligrams per liter.
Each circle represents a ground-water study

- Highest (greater than 5.0)
- Medium (0.5 to 5.0)
- Lowest (less than 0.5)

Background concentration

○ Bold outline indicates median values greater than background concentration (2 milligrams per liter)

Agricultural areas

Nitrate concentrations in agricultural areas were among the highest measured, but not all agricultural areas had median values above the national background concentration.

Average annual total nitrogen input—
in pounds per acre, by county, for 1991–94.
Inputs are from fertilizer, manure, and the atmosphere

- Highest (greater than 25)
- Medium (6 to 25)
- Lowest (less than 6)

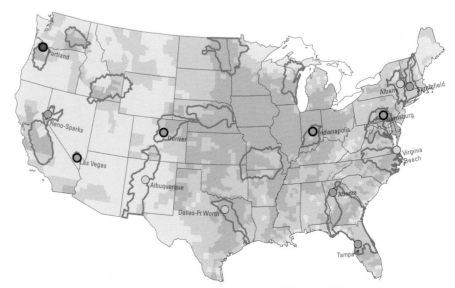

Urban areas

Nitrate concentrations in urban areas generally were lower than in agricultural areas, but 40 percent of urban areas had median values above the national background concentration.

See p. 31 for more information about these maps

NITROGEN IN STREAMS

Average annual concentrations of total nitrogen in about 50 percent of agricultural streams ranked among the highest of all streams sampled in the first 20 Study Units, and concentrations in about 36 percent of urban streams were among the highest measured. In contrast, total nitrogen levels in streams draining relatively undeveloped, forested watersheds (not shown on the national maps) ranked among the lowest of all streams sampled.

High concentrations of nitrogen in agricultural streams correlated with nitrogen inputs from fertilizers and manure used for crops and from livestock wastes. The Upper Midwest is a notable region of high nitrogen levels in agricultural streams; however, there are also many such examples in the West and East. High nitrogen levels in urban streams probably were related to nitrogen introduced from fertilizers applied to suburban lawns and golf courses, emissions from automobiles and electric powerplants, and effluent from sewage treatment facilities.

Streams and large rivers that drain areas of mixed land use had average annual concentrations of total nitrogen at various levels across the Nation, with no apparent regional pattern. The highest average annual concentrations occurred in watersheds with the highest nitrogen inputs and in rivers downstream from major metropolitan areas. The lowest concentrations in streams draining areas of mixed land use were for watersheds having considerable proportions of forest.

For all streams sampled, the highest concentrations of total nitrogen corresponded to broad patterns of nitrogen inputs from agricultural and urban areas. Coastal waters near such areas of high nitrogen input are at greatest risk of eutrophication.

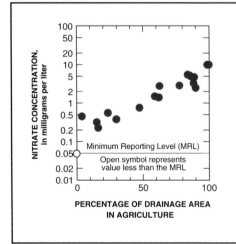

During high spring streamflows following fertilizer application in the northern Willamette Basin, nitrate concentrations increased in proportion to the percentage of drainage area in agriculture. Nutrient concentrations also were found to increase with the percentage of drainage area in agriculture for watersheds in the Ozark Plateaus, Potomac River Basin, and Trinity River Basin.

A national ranking of TOTAL NITROGEN concentrations in streams

Average annual concentration of total nitrogen—
in milligrams per liter

- ● Highest (greater than 2.9)
- ● Medium (0.6 to 2.9)
- ○ Lowest (less than 0.6)

Agricultural streams

Total nitrogen concentrations in agricultural streams were among the highest measured and generally correlated with nonpoint nitrogen inputs across the Nation.

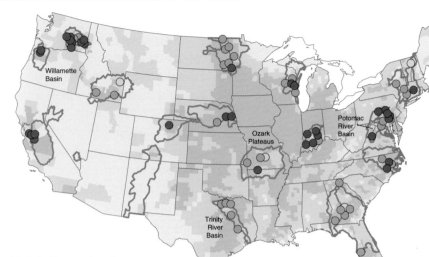

Average annual total nitrogen input—
in pounds per acre, by county, for 1991–94.
Inputs are from fertilizer, manure, and the atmosphere

- ▢ Highest (greater than 25)
- ▢ Medium (6 to 25)
- ▢ Lowest (less than 6)

Urban streams

The highest total nitrogen concentrations in urban streams typically were in densely populated areas in relatively arid Western basins and in the Northeast.

Rivers and streams with mixed land use

Total nitrogen concentrations generally correlated with nonpoint nitrogen inputs, but levels in large rivers downstream from major metropolitan areas were among the highest measured.

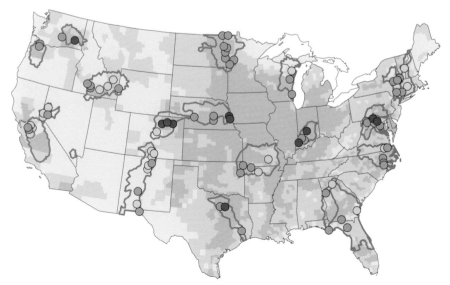

See p. 31 for more information about these maps

PHOSPHORUS IN STREAMS

In most streams draining agricultural, urban, or mixed land use, concentrations of total phosphorus were greater than background concentrations and the USEPA desired goal for preventing nuisance plant growth in streams. About one-half of urban streams had average annual concentrations of total phosphorus that ranked among the highest measured in the first 20 Study Units. The highest average annual concentrations of total phosphorus were in streams near metropolitan areas in the semiarid western and southwestern regions of the Nation. Examples include the Santa Fe River downstream from Santa Fe, New Mexico; Las Vegas Wash downstream from Las Vegas, Nevada; and the South Platte River downstream from Denver, Colorado. In these areas, discharges from wastewater treatment plants can be a significant proportion of the streamflow.

The broad geographical pattern observed for concentrations of total nitrogen in streams also holds true for concentrations of total phosphorus. This comparability is not surprising because proportions of nitrogen and phosphorus fertilizer used across the Nation are similar. Elevated phosphorus concentrations in agricultural streams can also come from livestock waste, such as in Prairie Creek in the Central Nebraska Basins, or from poultry wastes, such as in streams of the Apalachicola-Chattahoochee-Flint River Basin and the Ozark Plateaus. In addition, agricultural sites can be affected by effluent from upstream wastewater treatment plants, such as in Turlock Irrigation District Lateral 5 in the San Joaquin-Tulare Basins. Some high concentrations of phosphorus can occur naturally. For example, concentrations of phosphorus in the Pembina River of the Red River of the North Basin were high because most agricultural land in this area includes phosphorus-rich soils in relatively steep, easily eroded terrain; high concentrations of phosphorus in some streams in the Albemarle-Pamlico Drainage were derived from ground water in contact with phosphate minerals.

Concentrations of total nitrogen and total phosphorus typically were low in large rivers that drain areas of mixed land use. Examples include the Altamaha River, Georgia; Connecticut River, Connecticut; Menominee River, Wisconsin; and Upper Snake River, Idaho. In these large watersheds, streams draining forested and other relatively undeveloped land dilute nutrient-rich runoff from agricultural and urban areas. A few large rivers (such as the South Fork of the Palouse River, Washington, and the Trinity River, Texas) had extremely high average annual concentrations of total nitrogen and (or) total phosphorus that can be attributed primarily to the effect of upstream discharges of wastewater effluent. Large watersheds dominated by agricultural land, such as in the Great Plains and Upper Midwest, also exhibited concentrations of total phosphorus that ranked among the highest measured.

Kevin F. Dennehy

Effluent can make up a substantial part of the streamflow in some areas. For example, wastewater treatment plants annually contribute about 69 percent (and at times 100 percent) of the flow in the South Platte River downstream from Denver, Colorado. About 1,200 tons of phosphorus enter the South Platte River Basin every year from wastewater treatment plants.

Phosphorus transport in watersheds is complex. Phosphorus commonly attaches to soil particles and either remains close to application areas or moves to streams primarily by soil erosion. As phosphorus levels increase in some soils, a larger amount is available for transport in the dissolved form. This amount varies according to the phosphorus adsorption capacity of the soil. Areas with high levels of soil phosphorus relative to their soil adsorption capacity export relatively large amounts of phosphorus to streams.

A national ranking of TOTAL PHOSPHORUS concentrations in streams

Average annual concentration of total phosphorus— in milligrams per liter
- Highest (greater than 0.25)
- Medium (0.045 to 0.25)
- Lowest (less than 0.045)

Background concentration
O Bold outline indicates median values greater than background concentration (0.1 milligram per liter)

Agricultural streams

Total phosphorus concentrations in agricultural streams were among the highest measured and generally correlated with nonpoint phosphorus inputs across the Nation.

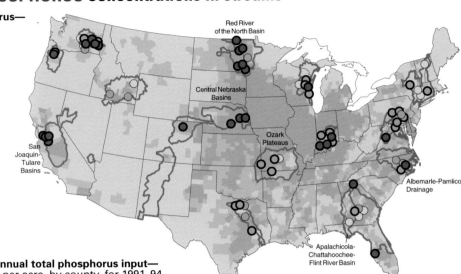

Average annual total phosphorus input— in pounds per acre, by county, for 1991–94. Inputs are from fertilizer and manure
- Highest (greater than 5)
- Medium (2 to 5)
- Lowest (less than 2)

Urban streams

The highest total phosphorus levels in urban streams typically were in densely populated areas in relatively arid Western basins and in the East.

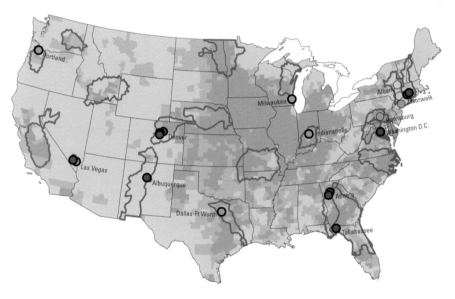

Rivers and streams with mixed land use

Total phosphorus concentrations generally correlated with nonpoint phosphorus inputs; however, in contrast to total nitrogen, levels in large rivers were highest in the Midwest, Great Plains, and West, where high concentrations of suspended sediment from erosion are common.

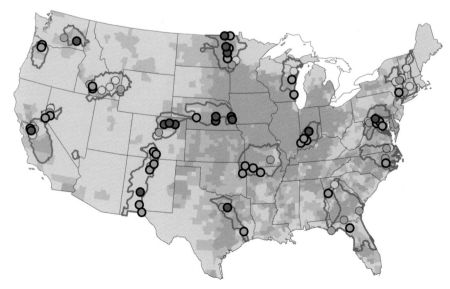

See p. 31 for more information about these maps

Differences in occurrence and behavior of nutrients complicate prediction of effects and management options

NUTRIENT INPUTS AND ENVIRONMENTAL FACTORS CONTROL NUTRIENT LOSSES FROM WATERSHEDS TO STREAMS

Enrichment of streams with nutrients is not simply explained by differences in land use. Land-management practices, nitrogen inputs to the land surface, local and regional environmental characteristics, and seasonal effects also control the degree of enrichment. Such an integration of factors explains why nutrient concentrations can be so different in different regions of the Nation, despite seemingly similar land-use settings.

Nutrient inputs to the land are key to explaining variations in the amount of nutrients lost from watersheds to streams (nutrient yields). The amounts of nutrients reaching streams generally increase as the total nonpoint nutrient inputs increase. For streams, nitrogen yields were less than or equal to about one-half of the total nonpoint inputs of nitrogen from the atmosphere, commercial fertilizer, and manure. This is consistent with the tendency of nitrate to dissolve in water and be transported with surface and subsurface runoff. Phosphorus yields, on the other hand, were less than or equal to about one-sixth of the total phosphorus inputs from commercial fertilizer and manure. Again, this is consistent with the general tendency of phosphorus to readily attach to soil particles rather than to dissolve in water that runs off to streams or seeps to ground water. In watersheds where crops and other plants cannot use all nutrients applied during a growing season, excess nutrients may be available for runoff to streams.

Local watershed characteristics and environmental settings also play key roles in determining nutrient yields. Watersheds with high nitrogen yields compared to total nitrogen inputs include Bachman Run, East Mahantango Creek, Kishacoquillas Creek, and Muddy Creek in the Lower Susquehanna River Basin; Broad Brook in the Connecticut, Housatonic, and Thames River Basins; and Zollner Creek in the Willamette Basin. These agricultural streams generally are in areas of high precipitation, which enhances runoff of surface water and flushing of shallow ground water, along with nitrogen, to streams. These watersheds also have a long history of farming, and they are located where soils and underlying geologic formations allow rapid movement of nitrogen-rich water through shallow aquifers and into streams.

Characteristics of soil drainage can accentuate or mitigate nutrient yields to streams. For example, the Lost River in the White River Basin exhibited high nitrate and phosphorus yields, particularly during high-flow conditions. Despite relatively low nutrient inputs, high nutrient yields probably result from sloping, clayey soils and shallow depth to permeable karst bedrock, which allow rapid transport of nutrients to the Lost River. Watersheds with high nitrogen inputs and low nitrogen yields, such as Prairie and Shell Creeks in the Central Nebraska Basins, have relatively flat-lying, sandy or silty

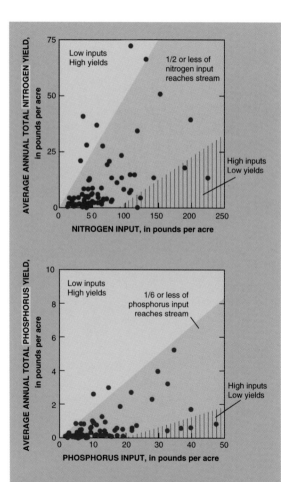

Average annual total nitrogen and total phosphorus yields to agricultural streams generally increase as nutrient inputs increase; however, a greater proportion of nitrogen than phosphorus input generally was lost to streams. The large amount of scatter in the data can be explained by local differences in agricultural practices, soils, geology, and hydrology across the Nation.

soils where water infiltrates readily and nitrate migrates to shallow ground water instead of being transported to streams.

Environmental factors controlling phosphorus yields to streams can be different from those controlling nitrogen yields. Watersheds with low phosphorus inputs and high phosphorus yields include Bullfrog Creek in the Georgia-Florida Coastal Plain—an unusual case reflecting contributions from naturally occurring phosphate minerals—and Broad Brook in the Connecticut, Housatonic, and Thames River Basins, which receives supplemental fertilization from phosphorus-rich manure. Basins with high phosphorus inputs and low yields include streams in the Lower Susquehanna River Basin, San Joaquin-Tulare Basins, and Albemarle-Pamlico Drainage. These streams gain significant flow from ground-water discharge, a source typically low in phosphorus because phosphates tend to be retained by the soil. Some of these same streams receive ground water that is high in nitrate because nitrogen inputs to the basins are high and nitrate can remain in solution. This is particularly true where denitrification, a microbial process that can transform nitrate to nitrogen gas, is not a controlling factor.

NITROGEN INPUTS AND ENVIRONMENTAL FACTORS CONTROL NITRATE CONCENTRATIONS IN SHALLOW GROUND WATER

Local differences in soils, geology, and hydrology affect nitrate migration from nonpoint sources to ground water in a more pronounced way than for nutrient yields to streams. Inputs of nitrogen were estimated from atmospheric, commercial fertilizer, and manure sources for areas within a one-third-mile radius of each monitoring well. Study areas with low inputs of nitrogen and high median nitrate concentrations (greater than about 4 mg/L) generally are underlain by karst or fractured rock or by unconsolidated sand and gravel that allow nitrate to move readily to shallow ground water. Such areas are found in the San Joaquin-Tulare Basins, Central Columbia Plateau, Red River of the North Basin, Western Lake Michigan Drainages, Lower Susquehanna River Basin, Potomac River Basin, and Connecticut, Housatonic, and Thames River Basins.

Areas with high nitrogen inputs but low median nitrate concentrations (less than about 2 mg/L) generally are underlain by relatively impermeable rock, silt, or clay, which impede downward movement of water. Examples of these areas are found in the Rio Grande Valley, White River Basin, and Western Lake Michigan Drainages. The Jerome-Gooding agricultural site in the Upper Snake River Basin also fell in the high-input and low-concentration group, but this was more likely related to the deep water table (median of 153 feet) in this area.

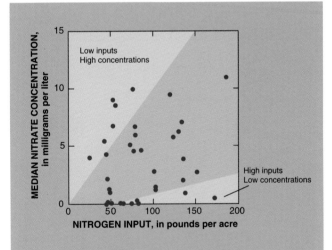

A plot of nitrogen inputs to agricultural land versus median nitrate concentrations in underlying shallow ground water shows considerable scatter. Porous soils and bedrock, which allow rapid downward movement of water and nitrate, underlie areas with low nitrogen inputs and high nitrate concentrations in shallow ground water. Areas with high inputs and low concentrations generally are underlain by less permeable geologic materials.

Nitrate concentrations in ground water generally were less than 0.1 mg/L in many parts of the White River Basin even though it receives some of the highest nitrogen fertilization rates in the Nation. The White River Basin is in the Upper Midwest, which contains some of the highest percentages of cropland (58 percent) and corn cropland (30 percent) in the United States. Nitrogen fertilization rates for corn often exceed 200 pounds per acre. Despite high nitrogen input rates, nitrate concentrations (shown in red, below) typically are low in ground water. This occurs because (1) poorly drained and impermeable glacial deposits, such as clay and silt, restrict the downward movement of water and nitrate to the water table, (2) nitrate is intercepted and transported to streams by tiles and ditches in many areas, and (3) nitrate is converted by denitrification to other forms of nitrogen where dissolved-oxygen concentrations in ground water are low. However, nitrate concentrations can be locally elevated in areas formerly traversed by glacial streams. The streams deposited coarse-textured and well-drained sand and gravel that allow rapid infiltration of water and enable nitrate to move below the root zone before it can be taken up by plants. Seventeen percent of shallow wells in these deposits had nitrate concentrations that exceeded the drinking-water standard of 10 mg/L.

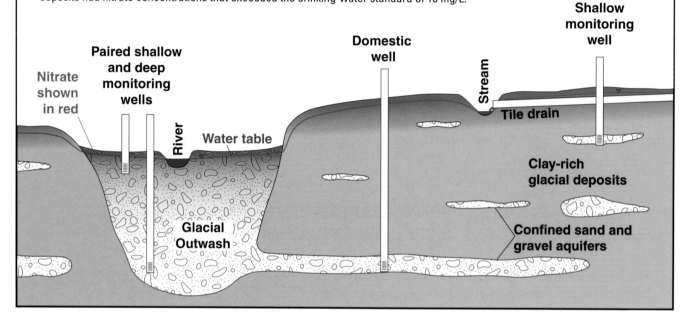

Nutrient concentrations vary seasonally

Nutrient concentrations vary throughout the year, largely in response to changes in precipitation and streamflow and to differences in time since fertilizer or manure application. Nutrient concentrations in streams typically are elevated during high spring and summer streamflows, or peak irrigation periods, following fertilizer application. In two agricultural streams in the Western Lake Michigan Drainages, for example, more phosphorus was transported during storms in June 1993 than during the 24 months that followed.

High nutrient concentrations also can be found in streams during seasonal low-flow conditions. Nitrate concentrations in agricultural streams can be high during winter low flow because of contributions from ground-water discharge and (or) because algal uptake is low. Nitrogen and phosphorus concentrations in streams downstream from metropolitan areas may be highest during various seasonal low flows, when contributions from point sources are greater relative to streamflow, and dilution is less.

Nitrate levels in shallow ground water can change throughout the year, but typically the seasonal changes are noticeable only in the upper 5-10 feet of the water table in surficial aquifers. For example, nitrate concentrations in shallow ground water from less than 10 feet below the water table in parts of the Red River of the North Basin ranged from about 8 to 25 mg/L from March 1994 through September 1995. This variation was related in part to the timing of important recharge periods, which generally occurred when spring snowmelt and major summer rainstorms coincided with irrigation periods, and in part to variations in the timing and application of fertilizers applied to crops.

Nitrate concentrations in shallow ground water beneath agricultural land can change seasonally in response to irrigation patterns.

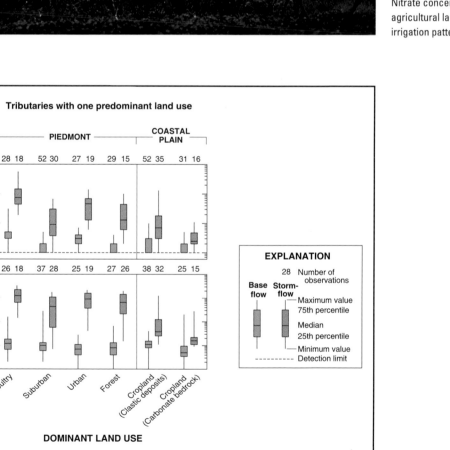

Total phosphorus concentrations in streams of the Apalachicola-Chattahoochee-Flint River Basin were higher during stormflow than during low flow and correlated with suspended sediment concentrations from March 1993 through September 1995

Stream-aquifer interactions control nitrate concentrations near some stream reaches

Irrigation and agricultural drainage can play a major role in the timing and magnitude of nutrient concentrations, particularly in the western part of the Nation, where large fluctuations in streamflow occur because of diversions for irrigation. Return flows from agricultural land during the irrigation season can account for most of the flow in many western streams and rivers, and concentrations of potential contaminants often are highest during peak irrigation periods. In addition, low nutrient concentrations in irrigation canals can dilute concentrations in ground water in areas where direct connections occur between the canals and adjacent aquifers.

Stream-aquifer interactions can affect nutrient concentrations differently during different times of the year in the same river reach. For example, nitrate concentrations in shallow ground water adjacent to the lower Suwannee River in the Georgia-Florida Coastal Plain vary seasonally because of a cycle of water exchange between the river and the adjoining aquifer. During summer low flow, ground water containing high nitrate concentrations enters the river, increasing river nitrate concentrations. During spring high flow, river water low in nitrate enters the aquifer, resulting in a decrease in ground-water nitrate concentrations adjacent to the river.

Stream-aquifer interactions also can affect nutrient concentrations differently in different parts of the same river basin. For example, nitrate concentrations in about one-half of the wells sampled near the South Platte River in Colorado exceeded the USEPA drinking-water standard. Ground water contributes a substantial amount of flow to the river in this area, but concentrations of nitrate in the river were substantially lower than in ground water because microbial denitrification removed nitrate as ground water passed through the streambed. Farther downstream in Nebraska, ground water in the alluvial aquifer adjacent to the Platte River is used for public supply by Nebraska's largest cities, including Omaha, Lincoln, Grand Island, and Kearney. Pumping water from wells in this aquifer induces flow of Platte River water into the aquifer and has the potential to decrease nitrate concentrations in the ground water.

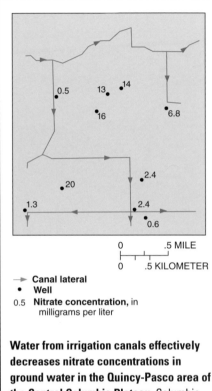

→ Canal lateral
• Well
0.5 Nitrate concentration, in milligrams per liter

Water from irrigation canals effectively decreases nitrate concentrations in ground water in the Quincy-Pasco area of the Central Columbia Plateau. Columbia River water diverted for irrigation leaks from canals and decreases nitrate concentrations by dilution in shallow ground water near the canals.

Modeling integrates information to estimate risks of nitrate contamination to shallow ground water

Models can integrate information on chemical use, land use, and environmental factors to help explain water-quality conditions over broad geographic regions. One USGS model, based on nationwide data, was developed to estimate the risk of nitrate contamination to shallow ground water across the United States.[30] The model integrates nitrogen inputs and aquifer vulnerability by use of Geographic Information System (GIS) technology. Nitrogen inputs include commercial fertilizer and manure application rates, atmospheric contributions, and population densities (the latter representing residential and urban nitrogen sources, such as septic systems, fertilizers, and domestic animal waste). Aquifer vulnerability is represented by soil-drainage characteristics—the ease with which water and chemicals can seep to ground water—and the extent to which woodlands are interspersed with cropland.

Nitrate concentrations measured in the first 20 Study Units generally conform to the national risk map. Nitrate concentrations are expected to be lowest in the areas shown in green, where nitrogen inputs and aquifer vulnerability are lowest, and highest in the areas mapped in red, which represent regions where nitrogen inputs and aquifer vulnerability are highest. **Anticipating where and what types of nitrate conditions exist can help focus regional or national water-management goals and monitoring strategies on the most vulnerable areas.**

Use of the risk map to identify and prioritize contamination at a more detailed level than presented here is not advised because local variations in land use, irrigation practices, aquifer type, and rainfall can result in nitrate concentrations that do not conform to risk patterns shown at the national scale.

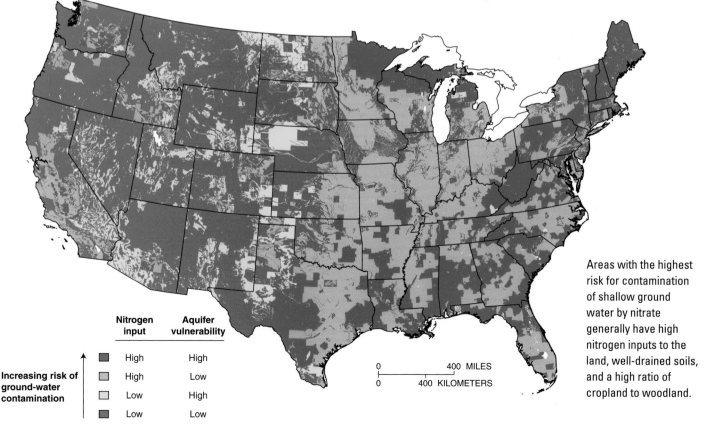

Increasing risk of ground-water contamination	Nitrogen input	Aquifer vulnerability
	High	High
	High	Low
	Low	High
	Low	Low

0   400 MILES
0   400 KILOMETERS

Areas with the highest risk for contamination of shallow ground water by nitrate generally have high nitrogen inputs to the land, well-drained soils, and a high ratio of cropland to woodland.

Nutrient conditions have changed over time in streams

Decades of monitoring may be necessary to adequately assess the effects of land- and water-management decisions on water quality. For example, decreases in phosphorus concentrations resulting from improved wastewater treatment technology and phosphate detergent bans have been documented in the Apalachicola-Chattahoochee-Flint River Basin; Albemarle-Pamlico Drainage; Connecticut, Housatonic, and Thames River Basins; Lower Susquehanna River Basin; Potomac River Basin, and Western Lake Michigan Drainages.

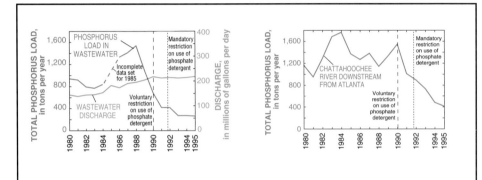

Improvements in wastewater treatment and bans on phosphate detergents have resulted in decreased phosphorus concentrations in the Chattahoochee River downstream from Metropolitan Atlanta. Wastewater discharge to the Chattahoochee River from the six largest Metropolitan Atlanta wastewater treatment facilities increased by about 50 percent from 1980 to 1995; however, the total phosphorus load from these facilities decreased by about 83 percent relative to the highest load recorded in 1988. Improvements in wastewater treatment account for about two-thirds of the decrease in phosphorus load, and restrictions on phosphate detergents account for about one-third of the decrease. By 1995, decreased phosphorus loads from point sources had resulted in total phosphorus loads in the Chattahoochee River downstream from Metropolitan Atlanta that were 77 percent less than the highest load measured in the Chattahoochee River in 1984.

Ammonia has decreased, but nitrate has increased, in the Trinity River downstream from Dallas, Texas. As a result of upgrades to wastewater treatment plants in the Dallas area, concentrations of ammonia plus organic nitrogen decreased about 95 percent from 1974 to 1991 at five sites on the Trinity River. Nitrate concentrations increased by a similar magnitude during the same period because the ammonia was converted to nitrate. The decrease in ammonia has led to an increase in dissolved oxygen, which reduces the threat of fish kills.

Despite decreases in ammonia and phosphorus in the Connecticut, Housatonic, and Thames River Basins, nutrients are still considered an environmental concern in Long Island Sound. Significant downward trends in total phosphorus concentrations were documented in 13 of 16 streams and rivers from 1980 to 1992 in the Connecticut, Housatonic, and Thames River Basins. The decreased phosphorus concentrations are likely due to improvements in wastewater treatment and to the elimination or reduction of phosphates in detergents. Ammonia decreased and nitrate increased during the same period, primarily as a result of the improved wastewater treatment processes, which convert ammonia to nitrate. Although improved treatment technology has enhanced surface-water quality in many parts of the Study Unit, the total amount of nutrients (particularly nitrogen) discharged to Long Island Sound is still considered an environmental concern. Excess nutrients continue to cause algal blooms, which decay and result in low dissolved-oxygen concentrations and poor habitat for fish and other marine animals in the Sound.

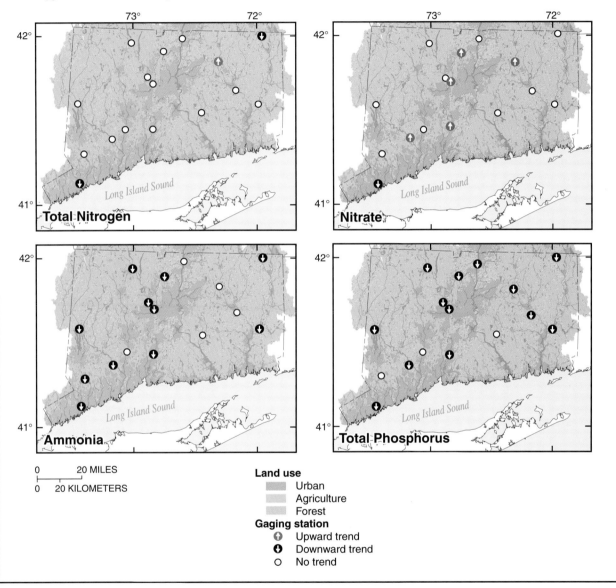

Removal of ammonia from point sources has enhanced stream quality in several Study Units, including the Connecticut, Housatonic, and Thames River Basins; Lower Susquehanna River Basin; Potomac River Basin; San Joaquin-Tulare Basins; and Trinity River Basin. Ammonia removal generally involves conversion to nitrate, and decreased ammonia concentrations typically have been accompanied by increased nitrate concentrations. Consequently, total nitrogen concentrations in these streams have remained about the same. Although toxicity to aquatic life has decreased as a result of ammonia removal, potential for eutrophication of surface waters probably has not changed.

Nutrient conditions have changed over time in ground water

Little information exists about trends of nitrate in ground water, particularly at a national scale, because few monitoring programs have been designed to look at the quality of ground water over time. Some information on nitrate trends is available, however, for the Upper Snake River Basin and San Joaquin-Tulare Basins. Studies in the San Joaquin Valley indicate that from 1950 to 1980, the largest source of nitrate (nitrogen fertilizer) increased from 114 to 745 million pounds per year. Concentrations of nitrate in ground water also increased, from less than 2 mg/L in the 1950s to about 5 mg/L in the 1980s.

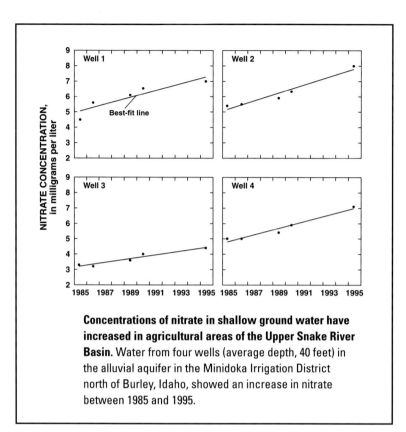

Concentrations of nitrate in shallow ground water have increased in agricultural areas of the Upper Snake River Basin. Water from four wells (average depth, 40 feet) in the alluvial aquifer in the Minidoka Irrigation District north of Burley, Idaho, showed an increase in nitrate between 1985 and 1995.

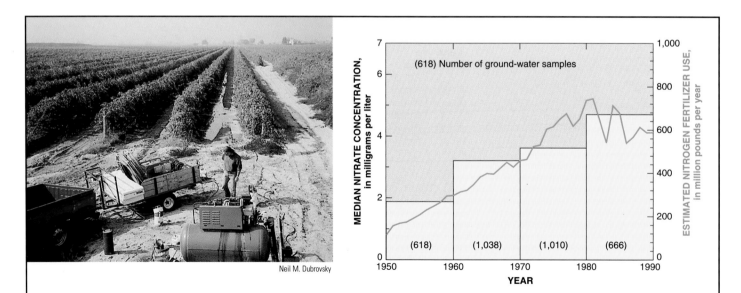

Neil M. Dubrovsky

Fertilizer use and nitrate concentrations in ground water in the eastern San Joaquin Valley (left) generally have increased over the last four decades. Although confined animal feeding operations and manure production also have increased during this period, nitrogen fertilizer is still considered to be the largest single source of nitrate to ground water.

The effects of past and present land-use practices may take decades to become apparent in ground water. When weighing management decisions for protection of ground-water quality, it is important to consider the time lag between application of nitrogen to the land and arrival of nitrate at a well. This time lag generally decreases with increasing aquifer permeability and with decreasing depth to water. In response to reductions in nitrogen applications to the land, the quality of shallow ground water will improve before the quality of deep ground water, which could take decades.

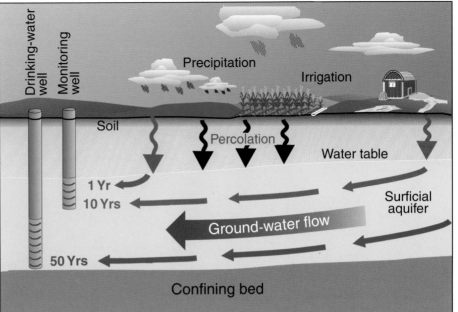

Nitrate concentrations have decreased in shallow ground water in parts of the Central Nebraska Basins. In the mid-1980s, the Central Platte Natural Resources District (CPNRD) established fertilizer management areas in part of the central Platte Valley, where nitrate concentrations were as high as 40 mg/L. Stringent guidelines were imposed on the timing and application rates of fertilizer in an area where the median nitrate concentration had increased from about 8 mg/L in 1974 to about 18 mg/L in 1986. In 1994, after implementation of the fertilizer management strategy, the median nitrate concentration decreased to less than 2 mg/L. It is important to note, however, that local variations in soil characteristics, amounts of recharge, and other factors affect responses to management strategies: nitrate concentrations in nearly 25 percent of the wells sampled by the CPNRD in the area with the most stringent guidelines continued to exceed 20 mg/L in 1994.

W.H. Mu

Kevin J. Breen

Pesticides used to control weeds, insects, and other pests on farms and in urban areas can be harmful to humans and the environment if they contaminate our water resources

Dennis A.

W.H. Mullins © 1974

Gregg Pat

Pesticides

Results of NAWQA studies show that pesticides are widespread in streams and ground water sampled within agricultural and urban areas of the Nation. As expected, the most heavily used compounds are found most often, occurring in geographic and seasonal patterns that mainly correspond to distributions of land use and associated pesticide use.

The frequency of pesticide contamination, however, is greater than expected. At least one pesticide was found in almost every water and fish sample collected from streams and in about one-half of all wells sampled. Moreover, individual pesticides seldom were found alone— almost every water and fish sample from streams and about one-half of samples from wells with a detected pesticide contained two or more pesticides.

For individual pesticides in drinking water, NAWQA results are generally good news relative to current water-quality standards and guidelines. Average concentrations in streams and wells rarely exceeded standards and guidelines established to protect human health. For aquatic life and wildlife, however, NAWQA results indicate a high potential for problems in many streams, particularly in urban areas, where concentrations of more than one pesticide often approached or exceeded established water-quality guidelines.

Important questions remain unanswered about potential risks of pesticide contamination to humans and the environment. Currently, standards and guidelines are available only for a limited number of individual pesticides, do not account for mixtures of pesticides or for pesticide breakdown products, and are based on tests that have assessed a limited range of potential health and ecological effects. Long-term exposure to low-level mixtures of pesticide compounds, punctuated with seasonal pulses of higher concentrations, is the most common pattern of exposure, but the effects of this pattern are not yet well understood.

The uncertainty about whether present-day levels of pesticide contamination are a threat to human health or the environment makes it imperative that we document and understand the nature of pesticide exposure, the causes of contamination, and the actions we can take to reduce pesticide levels in streams and ground water. ▶

Decades of pesticide use have resulted in their widespread occurrence in streams and ground water

More than 90 percent of water and fish samples from all streams contained one or, more often, several pesticides. Pesticides found in water were primarily those that are currently used, whereas those found in fish and sediment are organochlorine insecticides, such as DDT, that were heavily used decades ago. Most of the pesticides in use today are more water soluble and break down faster in the natural environment than the long-lived organochlorine insecticides of the past.[31]

About 50 percent of the wells sampled contained one or more pesticides, with the highest detection frequencies in shallow ground water beneath agricultural and urban areas and the lowest frequencies in major aquifers, which generally are deeper. Ground water has a lower incidence of pesticide contamination than streams because water infiltrating the land surface moves slowly through soil and rock formations on its way to ground water and through the aquifer. This contact with soil and rock and the slow rate of flow allow greater opportunity for sorption and degradation of pesticides, and varied flow pathways mean that some wells do not tap ground water that originated from places or times affected by pesticide use.

Although streams and rivers are more vulnerable than ground water to rapid and widespread contamination, ground-water contamination is extremely difficult to reverse because of the slow rate of ground-water flow. **Management practices that reduce the transport of pesticides to streams can yield rapid improvements in water quality. Ground water, on the other hand, will respond slowly to changing practices—sometimes taking many years or even decades to recover.**

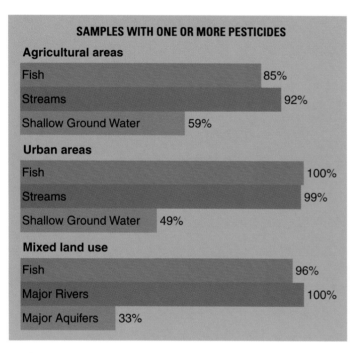

SAMPLES WITH ONE OR MORE PESTICIDES

Agricultural areas
Fish — 85%
Streams — 92%
Shallow Ground Water — 59%

Urban areas
Fish — 100%
Streams — 99%
Shallow Ground Water — 49%

Mixed land use
Fish — 96%
Major Rivers — 100%
Major Aquifers — 33%

WHAT WAS MEASURED...

Many of the Nation's most heavily used agricultural and urban pesticides were measured in the NAWQA Program. The 83 target compounds analyzed in water include 76 pesticides and 7 selected breakdown products and account for about 75 percent of the Nation's agricultural use of synthetic pesticides. They include 17 of the top 20 herbicides and 15 of the top 20 insecticides.

Historically used organochlorine insecticides, like DDT, were measured in bed sediment and fish, where they accumulate and persist for decades. The 32 organochlorine compounds analyzed in bed sediment or fish consist of 8 individual parent compounds, 1 individual breakdown product, and 7 groups of parent compounds plus related breakdown products or chemical impurities in the manufactured product. These compounds account for more than 90 percent of the Nation's historical use of organochlorine insecticides in agriculture.

WHAT WAS NOT...

Many important pesticide compounds were not measured because of analytical and budget constraints. The top 20 herbicides not measured were glyphosate (ranked 10), MSMA (14), and propazine (17). The top 20 insecticides not measured were cryolite (12), acephate (13), dimethoate (14), methomyl (15), and thiodicarb (18). Other pesticides not measured include inorganic pesticides, such as sulfur and copper, oil, and biological pesticides. Important omissions also include numerous pesticide breakdown products and carrier agents that may affect water quality.

Although NAWQA studies are targeting the broadest and most complete range of pesticides ever measured in a single assessment, these omissions are important to keep in mind and must temper conclusions.

Further information on pesticides measured is available via the World Wide Web at <http://water.usgs.gov/lookup/get?nawqapest>.

Pesticides are a potential concern for human health and aquatic life

Most pesticides are manufactured compounds that are designed to kill specific pests, such as weeds and insects. Many pesticides have the potential to harm nontarget organisms, especially if the organisms are exposed to high levels or for a long period of time. In the early 1960s, Rachel Carson's widely publicized book "Silent Spring"[32] described the ecological impacts of DDT and other pesticides. Concerns about the unintended effects of pesticides continue to this day, and evaluation of the risk to humans and the environment from present-day levels of pesticide exposure remains highly controversial.

A difficult aspect of evaluating potential effects of pesticides is determining what may occur as a result of varying types and durations of exposure. Exposure is complicated by pesticide mixtures, breakdown products, strong seasonal concentration pulses, and high concentrations during stormflows. In contrast, most toxicity assessments are based on controlled experiments with a single contaminant over a limited range of concentrations.

Although uncertainties remain, water-quality standards and guidelines have been developed for many pesticides in order to protect human health and aquatic life, and they are used in this report to signal potential problem areas. Concentrations that exceed a standard or guideline, however, may not be a problem at some sites. Conversely, the absence of an exceedance does not ensure that there is no problem.

Some people believe that any presence of pesticides in their drinking water is too much, whereas others feel that the standards and guidelines established for many of the major pesticides provide adequate protection. Which of these perspectives is closest to the truth remains unclear, but certainly the effects of common patterns of pesticide exposure found in NAWQA studies have not yet been fully evaluated.

The uncertainty in whether or not present-day levels of pesticide contamination are a threat to human health or aquatic life makes it imperative that we understand the nature of exposure, the causes of contamination, and the actions we can take to reduce pesticide levels in streams and ground water. Only by accurately characterizing the nature and causes of environmental exposure can we develop effective strategies to minimize exposure and reliably evaluate relations between exposure and effects.

STANDARDS AND GUIDELINES FOR PROTECTING WATER QUALITY

Water-quality standards and guidelines generally are maximum acceptable concentrations of pesticides for protecting humans, aquatic life, or wildlife. They are established by the United States and other nations, international organizations, and some States and tribes. For this report, precedence was given to standards and guidelines established by the USEPA and then to those established by Canada or the International Joint Commission for the Great Lakes, although some states may have different standards and guidelines that take priority for particular water bodies.[25]

Drinking-water standards or guidelines have been established for 43 of the 76 pesticides analyzed, and aquatic-life guidelines have been established for 28 of the 76 pesticides. Aquatic-life or wildlife guidelines are available for 8 of the 16 pesticides (compounds or groups) analyzed in bed sediment or fish.

Current standards and guidelines do not completely eliminate risks because: (1) values are not established for many pesticides, (2) mixtures and breakdown products are not considered, (3) the effects of seasonal exposure to high concentrations have not been evaluated, and (4) some types of potential effects, such as endocrine disruption and unique responses of sensitive individuals, have not yet been assessed.

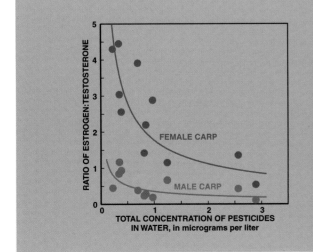

HORMONE LEVELS IN FISH SHOW SIGNS OF POSSIBLE ENDOCRINE DISRUPTION

A reconnaissance study of sex hormones in carp collected at 11 NAWQA stream sites indicates that pesticides may be affecting the ratio of estrogen to testosterone in both males and females.[33] The hormone ratio, which is sometimes used as an indicator of potential abnormalities in the endocrine system, was significantly lower at sites with the highest pesticide concentrations. Although the lower hormone ratios may not be associated with measurable effects on fish populations, they are a signal that further investigation is needed.

The most frequently detected pesticides are those most heavily used…now or in the past

Not surprisingly, the top 15 pesticide compounds found in water are among those with the highest current use. They include five of the most heavily used agricultural herbicides and one degradation product, five herbicides that are extensively used in urban areas, and four of the most commonly used insecticides.

The pesticide compounds found most often in fish and bed sediment are related to three major groups of insecticides that were heavily used in the 1960s. Organochlorine compounds related to DDT and dieldrin were widely used in both agricultural and urban areas, and chlordane was mainly used in urban areas.

PESTICIDE DETECTIONS

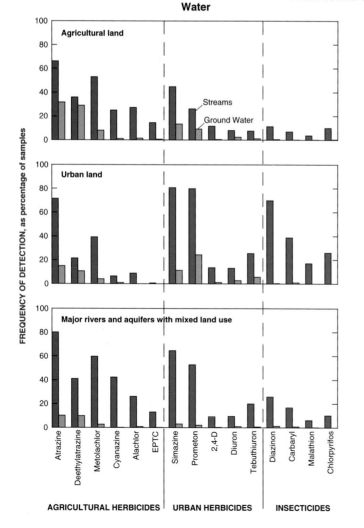

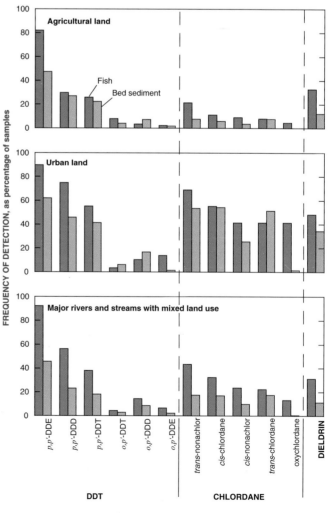

Different pesticides dominate in different land-use areas

The occurrence of pesticides in streams and ground water follows broad patterns in land use and associated pesticide use. The patterns are complex, however, and differ between streams and ground water because of the wide range of use practices and processes that govern the movement of pesticides in the hydrologic environment.

AGRICULTURAL AREAS

Herbicides are the most common type of pesticide found in streams and ground water within agricultural areas. The most common herbicides in agricultural streams were atrazine and its breakdown product

deethylatrazine (DEA), metolachlor, cyanazine, alachlor, and EPTC. All 5 of the parent compounds rank in the top 10 in national use. Atrazine was found in about two-thirds of all samples from agricultural streams, often occurring year-round.

Similar to streams, the most common compounds found in shallow ground water were atrazine and DEA, but only about one-third of the samples had detectable levels. The lower rates of atrazine and DEA detection in ground water compared to streams result from longer travel times, greater opportunity for

sorption or breakdown, and greater variability of source water in wells.

One of the most striking results for shallow ground water in agricultural areas, compared with streams, is the low rate of detection for several high-use herbicides other than atrazine. This is probably because these herbicides break down faster in the natural environment compared to atrazine. Studies show that break-down products of metolachlor, alachlor, and cyanazine are much more commonly found in ground water than are the parent compounds.[34]

Compared to herbicides, currently used insecticides were less frequently found in most agricultural streams. But some streams in agricultural areas with particularly high use of specific insecticides, such as diazinon in the San Joaquin-Tulare Basins, had among the highest concentrations measured. Insecticides were rarely detected in ground water in agricultural areas. **The less frequent occurrence of currently used insecticides in streams compared with herbicides, and their infrequent occurrence in ground water, result from their relatively low application rates and rapid breakdown in the environment.**

In contrast to currently used insecticides, the organochlorine insecticides of the past still persist in agricultural streams because of their extreme resistance to breakdown in the environment. DDT was the most

commonly detected organochlorine group—found in almost every fish sample— followed by dieldrin and chlordane. DDT and aldrin (which breaks down rapidly to dieldrin in the environment) were two of the top three insecticides used for agriculture in the 1960s.

URBAN AREAS

The most distinct differences between pesticides found in urban and agricultural areas are the greater prevalence of insecticides in urban streams and the relatively frequent occurrence of urban herbicides in both streams and shallow ground water. Insecticides were found more often, and usually at higher concentrations, in urban streams than in agricultural streams. Diazinon, carbaryl, chlorpyrifos, and malathion, which nationally rank 1, 8, 4, and 13 among insecticides used for homes and gardens, accounted for most detections in water. Historically used insecticides also were found more frequently in urban streams. Urban streams had the highest detection frequencies of DDT, chlordane, and

Peter E. Hughes

Kevin J. Breen

dieldrin in fish and bed sediment, and the highest concentrations of chlordane and dieldrin. Chlordane and aldrin were widely used for termite control until the mid-1980s, although their agricultural uses were restricted during the 1970s.[35, 36] Much more chlordane was used for termite control than for agriculture.

Insecticides in urban streams are a concern for aquatic life, for downstream water supplies, and possibly for recreational users. Effective management will likely require a combination of reducing current home, garden, and commercial use and controlling sediment sources to streams.

Similar to agricultural areas, insecticides were seldom detected in ground water in urban areas. Interestingly, however, the most commonly detected insecticide in shallow ground water was dieldrin, which was found in about 3 percent of the wells sampled. Although dieldrin is not very mobile in water, its environmental persistence and the heavy historical use of dieldrin and aldrin have combined to yield contamination of some wells.

The herbicides most commonly found in urban streams, in addition to atrazine and metolachlor, are simazine, prometon, 2,4-D, diuron, and tebuthiuron, all of which are commonly used in nonagricultural settings for maintenance of roadsides, commercial areas, lawns, and gardens. Prometon and 2,4-D have among the

highest frequencies of urban use. Of the urban herbicides, 2,4-D, simazine, and diuron also have substantial agricultural use, ranking in the top 25 nationally. Diuron and 2,4-D were not detected as frequently as other compounds with similar use, probably because the analytical method for these two compounds is less sensitive and resulted in fewer detections than for other compounds, even when concentrations were similar. As in streams, the most frequently found herbicides in shallow ground water in urban areas were atrazine, DEA, simazine, and prometon. Unlike streams, however, metolachlor was seldom detected, probably because of its lower urban use and lower persistence in the environment compared to the other herbicides.

Barbara J. Dawson

Peter A. Steeves

Pesticides found in major rivers and aquifers reflect contributions from both agricultural and urban areas

Major rivers and streams draining areas of mixed land use contain pesticides from both agricultural and urban sources and from both past and present use. In water that comes mainly from agricultural areas, the most commonly found pesticides are the major herbicides atrazine (and DEA), metolachlor, cyanazine, and alachlor. In water that comes mainly from urban areas, the most common pesticides are the herbicides simazine and prometon and the insecticides diazinon and carbaryl.

Like water, the fish and bed sediment of major rivers and streams with mixed land-use influences contain mixtures of organochlorine insecticides from agricultural and urban areas. Detection frequencies and concentrations of DDT, dieldrin, and chlordane were generally intermediate between those of agricultural and urban streams.

Many large rivers with mixed land-use influences tend to have lower concentrations of pesticides compared with agricultural and urban streams because of a larger influence of undeveloped land. Some rivers in intensive agricultural regions, however, have concentrations that are similar to those in agricultural streams, although they are less variable over time. Rivers with mixed land uses almost always contain detectable pesticides that reflect the diversity of sources present.

In contrast, ground water in major aquifers has a substantially lower frequency of pesticide occurrence than shallow ground water in agricultural and urban areas. This difference results from the generally deeper wells sampled in major aquifers and the greater influence of undeveloped areas. Additionally, owing to the slow rate of ground-

water flow, much of the water sampled in the major aquifers may have infiltrated into the ground before pesticides were applied. The two most frequently detected compounds in major aquifers were atrazine and DEA, resulting from the high and extensive use of atrazine, the greater extent of agricultural land compared to urban land affecting most major aquifers sampled, and the high mobility and long-lived nature of atrazine and DEA.

Because the pesticides found in major rivers and aquifers reflect contributions from both agricultural and urban land uses, efforts to improve the quality of these water resources will require management of nonpoint sources in both agricultural and urban areas.

This satellite image of the Central Columbia Plateau, taken in 1992, shows irrigated fields in green and fallow fields and rangeland in red. Agricultural runoff, tile drainage, and return flows from the irrigated farmland drains into the Columbia River, which forms the western border of the area before the Snake River joins it from the east.

Geographic distributions of pesticides follow patterns in land use and pesticide use

An essential step toward understanding and managing the effects of pesticides on water quality is to examine the geographic distribution of pesticide levels in relation to land use and pesticide use and to determine areas of the Nation and environmental settings that merit the greatest concern and attention.

The geographic distribution of pesticide levels is summarized in a series of maps that show results for herbicides and insecticides in streams and ground water for agricultural, urban, and mixed land uses, the latter including major rivers and aquifers. To identify potential water-quality problems, pesticide concentrations in water and bed sediment from streams are compared to aquatic-life guidelines because most streams sampled are not directly used as drinking-water sources. Pesticide concentrations in shallow ground water and water from major aquifers are compared to drinking-water standards and guidelines for human health. Most of the major aquifers, and shallow ground water in about one-half of the study areas, are sources of drinking water. Methods used to construct the maps are explained on page 31.

The national maps show national and regional patterns, or in some cases the apparent lack of pattern, in pesticide levels. They cannot, however, show important aspects of local variability in pesticide levels—for this, the reader is referred to the individual reports available for each NAWQA Study Unit (see page 80).

HERBICIDES EXCEEDED WATER-QUALITY STANDARDS OR GUIDELINES IN SOME STREAMS IN THE CORN BELT

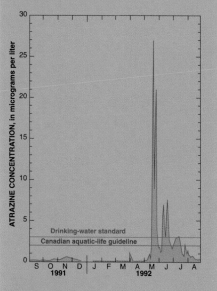

The heavy use of herbicides on corn in the Central Nebraska Basins is reflected in high atrazine concentrations in the Platte River during runoff from rainfall following spring herbicide applications. Low-level atrazine concentrations were found throughout much of the year, punctuated by seasonal pulses of high concentrations that exceeded the drinking-water standard (MCL) and the Canadian aquatic-life guideline. The annual average concentration, however, did not exceed the drinking-water standard.

Lincoln, Omaha, and smaller cities along the Platte River withdraw drinking water from an aquifer adjacent to the river. Much of the ground water that is pumped from the sand and gravel portions of this aquifer is vulnerable to contamination from atrazine in the Platte River. This is a concern to water providers

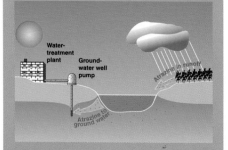

because studies have shown that conventional water treatment is ineffective in removing herbicides like atrazine from the treated water supplied to households.[38]

Herbicides in streams and major rivers were highest in the most intensively farmed agricultural regions

Total herbicide concentrations consistently ranked highest in agricultural streams and major rivers of the White River Basin and Central Nebraska Basins, which are on the eastern and western margins of the Corn Belt, respectively. The Corn Belt has the highest herbicide use in the Nation. The high concentrations measured in the White River Basin and Central Nebraska Basins are consistent with other studies in the Mississippi River Basin, which show broad-scale herbicide contamination of streams and rivers, including the Mississippi River.[37]

All seven agricultural streams and the two major rivers sampled in the White River Basin and Central Nebraska Basins frequently had concentrations of one or more herbicides that exceeded a Canadian aquatic-life guideline. Atrazine exceeded its guideline of 2 µg/L at all sites, and cyanazine exceeded its guideline of 2 µg/L at four sites. At this time, there are no national aquatic-life guidelines for these compounds in the United States, and individual States have varying guidelines.

Given the regional extent of intensive herbicide use and elevated levels of herbicides in streams within the Corn Belt, management strategies that are successful in reducing use and runoff of herbicides that are applied for corn and soybean production will likely lead to regional-scale improvements in water quality.

Other streams ranking high in herbicide concentrations were agricultural streams that drain intensively farmed areas in the Willamette Basin, San Joaquin-Tulare Basins, South Platte River Basin, and Trinity River Basin. A diverse group of herbicides, including trifluralin, metolachlor, and 2,4-D, in addition to atrazine and cyanazine, exceeded aquatic-life guidelines in one or more of these streams.

Most streams with low herbicide concentrations were agricultural streams in areas with low to moderate herbicide use in their drainage basins. Exceptions to this are low concentrations of herbicides in agricultural streams of the Red River of the North Basin and in the Southeast, even though use is moderate to high. One possible reason for the low concentrations in the Red River of the North Basin is a higher retention of herbicides in the soil because of particularly high levels of organic matter.

Among urban sites, only Las Vegas Wash in Las Vegas had relatively high herbicide concentrations compared to other streams. Only Little Buck Creek in the Indianapolis area had concentrations that exceeded a Canadian aquatic-life guideline, and that was in a small percentage of samples because of atrazine use on agricultural land in its watershed.

A national ranking of HERBICIDES in streams

Sum of herbicide concentrations
- ⬤ Highest 25 percent of streams
- ◐ Middle 50 percent
- ◯ Lowest 25 percent

Aquatic-life guidelines

○⁹ Bold outline indicates exceedance by one or more herbicides. Number is percentage of samples that exceeded a guideline

Agricultural streams

Concentrations were highest and most often exceeded aquatic-life guidelines in streams in the White River Basin and Central Nebraska Basins in the Corn Belt, where herbicide use is among the highest reported nationwide.

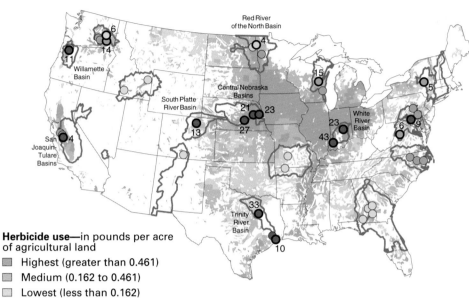

Herbicide use—in pounds per acre of agricultural land
- ⬛ Highest (greater than 0.461)
- ⬛ Medium (0.162 to 0.461)
- ⬜ Lowest (less than 0.162)
- ☐ No reported use

Urban streams

Most urban streams had moderate or low herbicide concentrations compared to streams in agricultural and mixed land-use settings.

Urban areas

Rivers and streams with mixed land use

Aquatic-life guidelines were exceeded in about one fourth of the samples from the two major rivers sampled in the Corn Belt, but most major rivers had moderate herbicide concentrations.

See p. 31 for more information about these maps

Insecticides in streams were highest in urban areas

Most urban streams sampled, plus two major rivers dominated by urban influences— the South Platte River downstream from Denver and the Trinity River downstream from Dallas-Fort Worth—had among the highest insecticide concentrations of all streams and rivers sampled. Nine of 11 urban streams and both rivers had concentrations that exceeded aquatic-life guidelines, usually in more than 20 percent of the samples. The most common insecticides to exceed guidelines were diazinon, chlorpyrifos, and malathion. Chlorpyrifos and malathion have USEPA aquatic-life criteria of 0.041 µg/L and 0.100 µg/L, respectively, and diazinon has a guideline of 0.080 µg/L established by the International Joint Commission for the Great Lakes.

Insecticides in urban streams, largely from use around homes and in gardens, parks, and commercial areas, frequently occur at levels of concern for aquatic life and may be a significant obstacle for restoring urban streams.

Most agricultural streams had moderate or low concentrations of insecticides but, as for herbicides, several streams that drain intensively farmed areas that are irrigated had among the highest insecticide levels. Although concentrations of insecticides in agricultural streams tended to be low compared to urban streams, concentrations above aquatic-life guidelines were common. For about one-half of the agricultural streams, samples exceeded a guideline for one or more insecticides. In addition to diazinon and chlorpyrifos, an insecticide that frequently exceeded its guideline in agricultural streams was methyl azinphos, which has a USEPA aquatic-life criterion of 0.010 µg/L.

Insecticide concentrations in most major rivers usually were lower than those measured in urban streams and exceeded aquatic-life guidelines in relatively few samples. Exceptions are the San Joaquin River, which drains farmlands with some of the heaviest insecticide use in the Nation, and the South Platte and Trinity Rivers, which are affected by both point and nonpoint sources from urban areas.

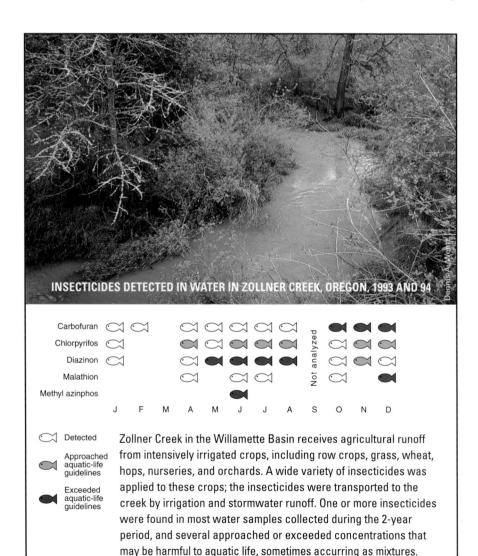

INSECTICIDES DETECTED IN WATER IN ZOLLNER CREEK, OREGON, 1993 AND 94

Carbofuran
Chlorpyrifos
Diazinon
Malathion
Methyl azinphos

Not analyzed

J F M A M J J A S O N D

Detected

Approached aquatic-life guidelines

Exceeded aquatic-life guidelines

Zollner Creek in the Willamette Basin receives agricultural runoff from intensively irrigated crops, including row crops, grass, wheat, hops, nurseries, and orchards. A wide variety of insecticides was applied to these crops; the insecticides were transported to the creek by irrigation and stormwater runoff. One or more insecticides were found in most water samples collected during the 2-year period, and several approached or exceeded concentrations that may be harmful to aquatic life, sometimes occurring as mixtures.

A national ranking of INSECTICIDES in streams

Sum of insecticide concentrations

- ● Highest 25 percent
- ● Middle 50 percent
- ○ Lowest 25 percent

Aquatic-life guidelines

○⁹ Bold outline indicates exceedance by one or more insecticides. Number is percentage of samples that exceeded a guideline

Agricultural streams

Most streams had moderate or low concentrations, but several in irrigated areas of the West had among the highest concentrations. About one-half of the agricultural streams had concentrations that exeeded an aquatic-life guideline.

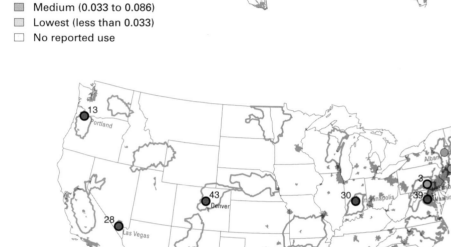

Insecticide use—in pounds per acre of agricultural land

- ■ Highest (greater than 0.086)
- ■ Medium (0.033 to 0.086)
- □ Lowest (less than 0.033)
- □ No reported use

Urban streams

Most streams had among the highest concentrations. Typically, 10 to 40 percent of samples had concentrations that exceeded one or more aquatic-life guidelines.

Rivers and streams with mixed land use

Concentrations were low to moderate except for the urban-affected South Platte and Trinity Rivers, and the San Joaquin River, which drains farmlands with some of the most intensive insecticide use in the Nation.

See p. 31 for more information about these maps

CONTROL OF SOIL EROSION IS KEY TO REDUCING ORGANOCHLORINE INSECTICIDES

Organochlorine insecticides bind strongly to soils and are carried with eroded soils to streams by runoff from irrigation and rainfall. In streams, the soil-bound insecticides may dissolve in water, remain suspended, or settle to the streambed. They also accumulate in fish. Under-standing and managing soil erosion is a key to reducing organochlorine contamination.

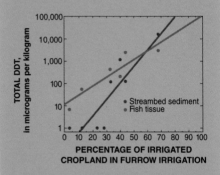

For example, furrow irrigation causes more erosion than sprinkler or drip irrigation. In the Central Columbia Plateau, DDT concentrations in streambed sediment and fish increased as the percentage of furrow irrigation in the basin increased.

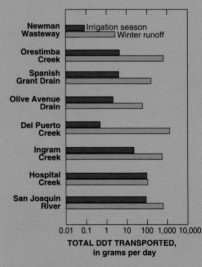

In the San Joaquin-Tulare Basins, the amount of DDT transported with suspended sediment in the San Joaquin River and tributaries generally was greater during winter runoff than during the irrigation season. Controlling irrigation-induced soil erosion would reduce but not eliminate DDT in the streams because large quantities are transported during infrequent storms.

Organochlorine insecticides were highest in urban streams and where historical agricultural use was greatest

Concentrations of organochlorine insecticides in bed sediment and fish correspond to land use and past application rates. Although most uses of organochlorine insecticides ended 10 to 25 years ago, they remain a significant water-quality issue for many streams. Overall, 14 percent of bed-sediment samples had concentrations that exceeded sediment-quality guidelines for protection of aquatic life,[39] and 19 percent of sites had concentrations in fish that exceeded New York guidelines for protection of fish-eating wildlife.[40] Compounds that most often exceeded guidelines were DDT and chlordane in bed sediment and DDT and dieldrin in fish.

Almost all urban streams had high or medium concentrations of the organochlorine insecticides compared with other sites. Sediment-quality guidelines were exceeded at 37 percent of urban sites, with several sites each in urbanized areas of the Connecticut, Housatonic, and Thames River Basins, Hudson River Basin, Trinity River Basin, and Georgia-Florida Coastal Plain. Concentrations in whole fish exceeded guidelines for the protection of fish-eating wildlife at 21 percent of urban sites.

In agricultural streams, concentrations of organochlorine insecticides were highest in areas of high past use. High concentrations were most common for streams in the Central Columbia Plateau, Georgia-Florida Coastal Plain, and Trinity River Basin. One or more sediment-quality guidelines were exceeded at 15 percent of agricultural sites, and concentrations in whole fish exceeded wildlife guidelines at 20 percent of sites.

Many streams and rivers with mixed land-use influences also had high concentrations in bed sediment, particularly in basins with extensive agricultural areas where past use was high, such as in the Southeast and the irrigated West, and in basins with high population density, such as in the Northeast. Sediment-quality guidelines were exceeded at 11 percent of these sites, and wildlife guidelines were exceeded in whole fish at 24 percent of these sites. In undeveloped areas, organochlorine concentrations generally were low and did not exceed sediment-quality guidelines.

A significant health concern in some regions is consumption of fish with high levels of organochlorine insecticides in their flesh. Human-health guidelines for edible fish tissue[41] are not directly applicable to NAWQA results, which are based on whole-fish analysis of mostly carp and suckers. Nevertheless, the NAWQA fish data provide a relative indication of potential concern. **At about 30 percent of NAWQA sites, insecticide concentrations in whole fish exceeded human-health guidelines for edible fish tissue.[41] For any of these streams that are active fisheries, additional assessment of fillets of edible species is advisable if this has not already been done.**

A national ranking of ORGANOCHLORINES in bed sediment

Sum of organochlorine insecticide concentrations
- Highest 25 percent
- Second highest 25 percent
- Lowest 50 percent (all with no detections)

Aquatic-life guidelines
- ○ Bold outline indicates exceedance by one or more organochlorine insecticides

Agricultural streams

Highest concentrations occurred where historical use was highest on crops such as cotton, peanuts, orchards, and vegetables.

Historical organochlorine use— in pounds per acre of agricultural land
- Highest (greater than 0.278)
- Medium (0.095 to 0.278)
- Lowest (less than 0.095)
- No reported use

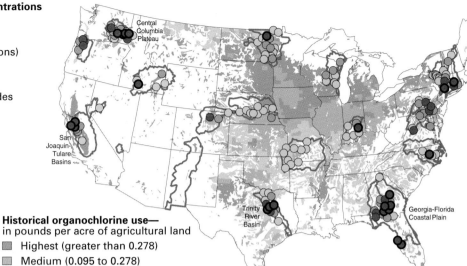

Urban streams

Most urban streams had higher concentrations than the majority of agricultural streams, and concentrations exceeded sediment-quality guidelines at almost 40 percent of the sites.

⚑ **Urban areas**

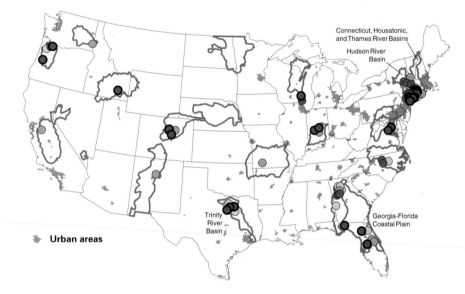

Rivers and streams with mixed land use

Concentrations followed the patterns in contributing agricultural and urban areas, with the highest concentrations in areas of high population densities or intensive historical use in agriculture.

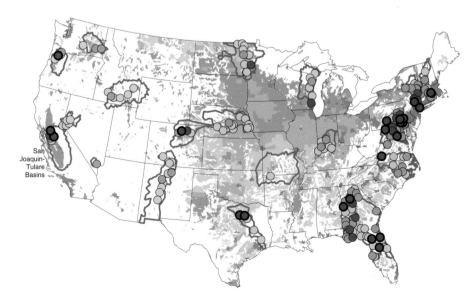

See p. 31 for more information about these maps

Herbicides in shallow ground water were most common beneath agricultural areas

The highest frequencies of detection for pesticides in ground water were for herbicides in shallow ground water beneath agricultural areas. In these areas where herbicide use was moderate to high, soil and geologic conditions favored rapid movement of herbicides to the ground water. Most studies of shallow ground water in agricultural areas detected herbicides in more than 50 percent of wells sampled.

Compared to streams, ground-water detections were dominated by fewer compounds—mainly those that have the combination of relatively high mobility and chemical stability that allows them to move and persist in the flow system long enough to reach a well. Only atrazine, its breakdown product DEA, metolachlor, prometon, and simazine were found in more than 5 percent of all wells.

Of the 36 studies of shallow ground water in agricultural areas, which included more than 1,000 wells, only one well in an unused shallow ground-water area in the Connecticut, Housatonic, and Thames River Basins had an atrazine concentration that exceeded the drinking-water standard of 3 µg/L.

Herbicides were moderately common in shallow ground water beneath urban areas. In an urban area of the Albemarle-Pamlico Drainage, a shallow aquifer used for drinking-water supply had one monitoring well where an atrazine concentration exceeded the drinking-water standard.

Major aquifers, all of which are drinking-water sources, are generally deeper than the shallow ground water studied and had distinctly lower detection frequencies of herbicides. Only 3 of 33 aquifers sampled had among the highest ranked detection frequencies, and none of the wells sampled in major aquifers had herbicide concentrations that exceeded drinking-water standards or guidelines.

Ground-water contamination, compared to stream contamination, is more strongly governed by soil and geologic conditions, and each well is uniquely affected by sources of pesticides and flow conditions in its immediate vicinity. Local variability

in these conditions can result in degradation of water quality in one or a few wells, even if most wells are not affected. The greatest frequencies of herbicide detection in major aquifers occurred in vulnerable settings. The three aquifers with the highest frequencies of detection were (1) the Platte River Alluvial aquifer in the Central Nebraska Basins, which is shallow and overlain by permeable sandy soils, (2) the Upper Floridan aquifer in the Appalachicola-Chattahoochee-Flint River Basin, which is a limestone formation where flow rates are high, and (3) a shallow limestone aquifer in the Lower Susquehanna River Basin.

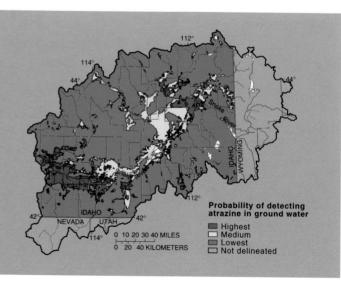

PREDICTING ATRAZINE CONTAMINATION IN GROUND WATER

As part of Idaho's State Pesticide Management plans for herbicides, maps have been developed to portray the potential for atrazine contamination in ground water in southeastern Idaho.[42] Atrazine data from the NAWQA Program in the Upper Snake River Basin were used to calibrate and verify predictive models. Significant factors used to successfully predict atrazine concentrations in ground water were atrazine use, land use, precipitation, soil type, and depth to ground water. Continued development of these types of modeling tools will aid in designing cost-effective programs for monitoring and protecting ground-water resources across the Nation.

Probability of detecting atrazine in ground water
- Highest
- Medium
- Lowest
- Not delineated

A national ranking of HERBICIDES in ground water

Herbicide detection frequency—Each circle represents a ground-water study
- ● Highest 25 percent
- ● Middle 50 percent
- ○ Lowest 25 percent

Drinking-water standards or guidelines
- ○³ Bold outline indicates exceedance by one or more herbicides. Number is percentage of wells that exceeded a a standard or guideline

Shallow ground water in agricultural areas

The highest detection frequencies occurred where use is moderate to high and where soil and geologic conditions promote rapid infiltration.

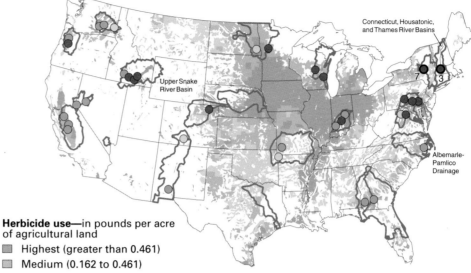

Herbicide use—in pounds per acre of agricultural land
- ■ Highest (greater than 0.461)
- ■ Medium (0.162 to 0.461)
- ■ Lowest (less than 0.162)
- □ No reported use

Shallow ground water in urban areas

Only two urban areas had detection frequencies in the highest 25 percent of all ground-water studies.

🦐 **Urban areas**

Major aquifers

Detections were infrequent, except for a few aquifers in vulnerable settings—shallow aquifers with permeable sandy soils or limestone formations.

See p. 31 for more information about these maps

Insecticides were seldom found in ground water but may be a concern in some areas

Insecticides, in contrast to herbicides, were not detected in a number of ground-water studies and, where detected, were usually found in less than 10 percent of wells. The most frequently detected insecticides in ground water were dieldrin and diazinon, although each was found in only 1 to 2 percent of all wells. The relative abundance of dieldrin was unexpected because of its low mobility in water compared with many currently used pesticides. Dieldrin, however, is one of the more mobile compounds within the historically used organochlorine group. Moreover, it is long-lived in the environment, which results in its great persistence in the ground-water flow system.

Although insecticides were much less common than herbicides in ground water, they exceeded drinking-water standards or guidelines more often. In all but one well where exceedances occurred, dieldrin was the insecticide that exceeded the guideline. The guideline used for dieldrin is a USEPA Risk Specific Dose of 0.02 µg/L, which corresponds to a cancer risk level of 1 in 100,000. The wells that exceeded the Risk Specific Dose for dieldrin were mainly wells tapping shallow ground water that is not used for human consumption.

The infrequent but potentially important occurrences of dieldrin in some wells may be the result of local contamination of individual wells. The combination of relatively shallow ground water and pesticide use in the vicinity of wells increases the likelihood that some wells will have flow pathways that allow pesticides to move from the land surface to the well, sometimes down the borehole itself.

Daniel J. Hippe

DIELDRIN PERSISTS IN SHALLOW URBAN GROUND WATER

In the Apalachicola-Chattahoochee-Flint River Basin, insecticide concentrations in ground-water samples generally were less than current drinking-water standards or guidelines. However, dieldrin concentrations in water samples collected during 1994–95 from 5 of 37 shallow wells and springs in Metropolitan Atlanta exceeded the USEPA Risk Specific Dose of 0.02 µg/L, which corresponds to a cancer risk level of 1 in 100,000. Dieldrin and aldrin, which breaks down to dieldrin in the environment, had been used on agricultural land prior to 1975 and for structural termite control until 1987.[36] Although this ground water is not used as a source of drinking water, the presence of dieldrin in ground-water samples collected several years after being banned is indicative of the compound's persistence in soils and ground water and its potential to be a problem in some wells.

A national ranking of INSECTICIDES in ground water

Insecticide detection frequency—Each circle represents a ground-water study or major aquifer

● Highest 25 percent
◐ Middle 46 percent
○ Lowest 29 percent (all with no detections)

Drinking-water standards or guidelines

○⁵ Bold outline indicates exceedance by one or more insecticides. Number is percentage of wells that exceeded a a standard or guideline

Shallow ground water in agricultural areas

Detection frequencies ranked low to moderate in most studies.

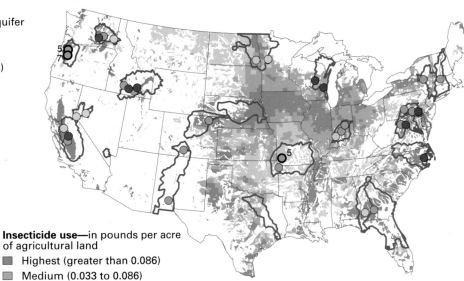

Insecticide use—in pounds per acre of agricultural land

■ Highest (greater than 0.086)
▦ Medium (0.033 to 0.086)
▢ Lowest (less than 0.033)
□ No reported use

Shallow ground water in urban areas

Detection frequencies ranked high in urban areas compared with other study areas but still were low compared to herbicides. Although aldrin and dieldrin have not been used for years, dieldrin was the most frequently detected insecticide.

✦ **Urban areas**

Major aquifers

Most major aquifers ranked low to moderate in detection frequency and, only one well exceeded a drinking-water standard or guideline (dieldrin).

See p. 31 for more information about these maps

Differences in occurrence and behavior of pesticides complicate evaluation of potential effects

PESTICIDES USUALLY OCCUR AS MIXTURES

Pesticides usually occur in mixtures of several compounds rather than individually, but most of our experience and research on environmental effects is based on exposure to individual compounds. **Therefore, it is vital that we understand and document the occurrence and composition of common low-level mixtures and begin to evaluate their effects.**

More than 50 percent of all stream samples contained five or more pesticides, and nearly 25 percent of ground-water samples contained two or more pesticides. In the Central Columbia Plateau, for example, 66 percent of ground-water samples with detections contained more than one pesticide, most commonly in shallow monitoring wells. The most common mixtures were found more than twice as frequently in streams than in ground water, except for the atrazine-DEA combination.

Mixtures of currently used pesticides in stream water may occur in combination with mixtures of organochlorine insecticides in bed sediment and fish. Moreover, about 50 percent of bed-sediment and fish samples with pesticide detections contained compounds from two or more of the major organochlorine groups.

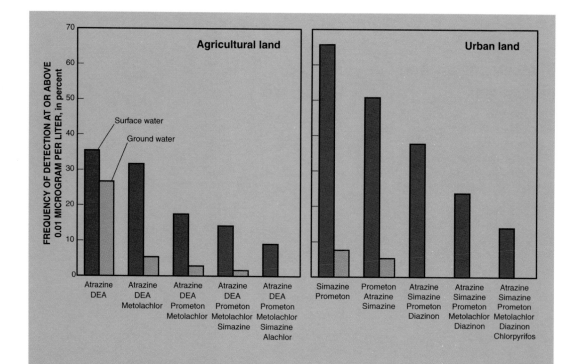

COMMON PESTICIDE MIXTURES IN WATER

The composition of the most common pesticide mixtures differs between urban and agricultural areas and between agricultural areas with different crops and pests. In urban areas, simazine and prometon were the most common pesticides found together, whereas atrazine, DEA (deethyl-atrazine), and metolachlor were the most common compounds found in mixtures from agricultural areas. Mixtures containing both herbicides and insecticides were a common occurrence in urban streams. More than 10 percent of urban stream samples contained a mixture of at least 4 herbicides plus the insecticides diazinon and chlorpyrifos.

BREAKDOWN PRODUCTS CAN BE IMPORTANT

Once released into the environment, pesticides undergo a series of chemical and biological reactions whereby the original pesticide breaks down into intermediate compounds, and eventually into carbon dioxide and other harmless compounds. Some breakdown products are short-lived, whereas others persist for years or decades. Little is known about the occurrence of many pesticide breakdown products, and even less is known about their effects on human health and aquatic life.

Of the thousands of possible breakdown products, few have been looked for in streams or ground water.[6,43] Some are less toxic than their parent compounds, whereas others have been found to have similar or even greater toxicities.

Only seven breakdown products were analyzed in water samples from the first 20 Study Units: 2,6-diethyl-aniline (parent pesticide, alachlor), 3-hydroxy-carbofuran (carbofuran), aldicarb sulfone and aldicarb sulfoxide (aldicarb), DDE (DDT), alpha-HCH (lindane), and DEA (atrazine). Of the parent pesticides, atrazine is the most heavily used, and both it and DEA were widespread in streams and ground water across the Nation. The two were found together

in about 35 percent of stream samples and about 25 percent of ground-water samples from agricultural areas.

With few exceptions, most of the other breakdown products were found in fewer than 1 percent of samples in each of the Study Units. However, several breakdown products of alachlor and metolachlor have been frequently found in other studies, often at much higher concentrations than the parent pesticide.[34, 44] As NAWQA evolves, more complete analyses of breakdown products are being added as analytical methods and budget constraints allow.

CONCENTRATIONS IN STREAMS FOLLOW STRONG SEASONAL PATTERNS

Seasonal patterns in concentrations and occurrences of pesticides in agricultural streams, which tend to repeat each year, correspond to patterns in use and streamflow, including contributions from ground water. Generally, the number and concentrations of herbicides found in most agricultural streams were highest from April through July, whereas insecticides occurred more variably throughout the summer. The spring herbicide pulse was commonly observed in corn-growing areas and other agricultural areas shortly after

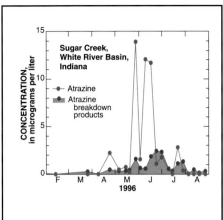

ATRAZINE AND ITS BREAKDOWN PRODUCTS WERE DETECTED THROUGHOUT THE YEAR IN THE WHITE RIVER BASIN

herbicide application, when herbicides were transported to streams in runoff induced by spring rain and irrigation. In some parts of the Nation, other patterns can occur. For example, some insecticides, such as diazinon in the San Joaquin-Tulare Basins, have patterns of high concentrations during the winter, resulting from the use of dormant sprays on orchards. Differences in patterns also may result from local water-management practices, including the timing of reservoir storage and water use, the timing of runoff from agricultural fields due to irrigation or storms, or ground-water contributions during periods of low streamflow. **Seasonal patterns need to be characterized and understood because they dictate the timing of high concentrations in drinking-water supplies and the times when aquatic organisms may be exposed to high concentrations during critical stages of their life cycle.** For example, some water suppliers reduce their use of certain surface-water supplies during spring runoff.

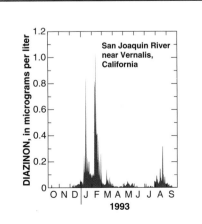

HIGH DIAZINON CONCENTRATIONS IN THE SAN JOAQUIN RIVER WERE COMMON FOLLOWING WINTER APPLICATION

Diazinon concentrations in the San Joaquin River near Vernalis, California, exceeded concentrations shown to be toxic to aquatic life during January and February 1993—following the winter application of diazinon, a dormant spray applied to control wood-boring insects in almond orchards in the San Joaquin-Tulare Basins.

Trends in pesticide concentrations follow changes in use

Pesticides in streams and ground water change over time as the types and amounts of chemicals in use change. With the exception of organochlorine insecticides, however, consistent data that are adequate for assessing long-term trends have not been widely collected. Examples for the organochlorine insecticides and recent changes in herbicide use illustrate the importance of tracking such trends.

ORGANOCHLORINE INSECTICIDES HAVE DECREASED

A striking historical trend is the reduction in concentrations of organochlorine insecticides in sediment and fish following restrictions on their use, yet they continue to occur at levels of concern. This trend is evident in sediment cores from lakes and reservoirs and by comparison of NAWQA findings to historical concentrations in fish measured by the U.S. Fish and Wildlife Service (USFWS).

As sediment erodes from the land surface over time, it is deposited in layers on the bottom of lakes and reservoirs. Age-dated sediment cores that penetrate these layered deposits can be used to track trends in sediment-associated contaminants within the drainage basin. Concentrations of total DDT (DDT plus breakdown products DDE and DDD) in sediment cores from lakes and reservoirs reflect high historical use of DDT followed by a ban in 1972. DDT concentrations peaked during the 1960s, which coincides with its peak use as an insecticide. Total DDT concentrations in sediment have decreased since 1972 in all sampled lakes and reservoirs that drain urban and agricultural areas within the United States.[45]

Unlike DDT, aldrin and chlordane were used for termite control until the late 1980s, long after their agricultural uses were cancelled in the early 1970s. Chlordane and dieldrin concentrations peaked in many agricultural areas during the 1970s, and decreased thereafter. In some urban lakes and reservoirs, however, such as White Rock Lake in the Trinity River Basin, chlordane and dieldrin peaked much later, probably as a result of continued urban use during the 1980s. This watershed is dominated by new (post-1960) urbanization.[46]

Concentrations of DDT, chlordane, and dieldrin in whole fish have declined nationally since the 1970s. To assess trends in DDT concentrations, NAWQA data for streams and rivers with mixed land influences were compared with similar data from 1969 to 1986 collected by the USFWS National Contaminant Biomonitoring Program.[47] Total DDT concentrations in fish declined markedly from 1969 to the present. The declines were greatest during the early 1970s, with concentrations since the mid-1980s showing a slower decline or even a plateau.

Despite the observed national decline in total DDT concentrations, the detection frequency for total DDT in whole fish from major rivers remains high (94 percent in the 1990s), and locally contaminated areas persist. This is probably caused by the presence of total DDT in the streambed and continued inputs of total DDT to hydrologic systems as contaminated soils erode into streams.

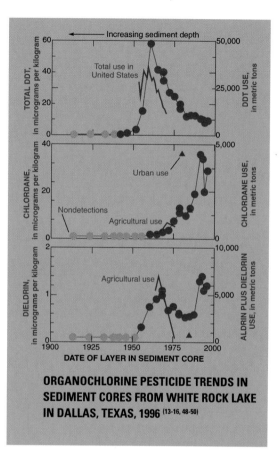

ORGANOCHLORINE PESTICIDE TRENDS IN SEDIMENT CORES FROM WHITE ROCK LAKE IN DALLAS, TEXAS, 1996 [13-16, 48-50]

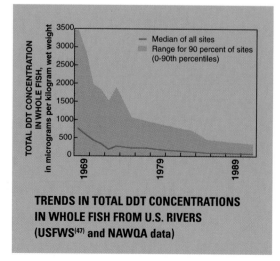

TRENDS IN TOTAL DDT CONCENTRATIONS IN WHOLE FISH FROM U.S. RIVERS (USFWS[47] **and NAWQA data)**

RECENT CHANGES IN HERBICIDE USE HAVE BEEN RAPIDLY REFLECTED IN STREAMS

Few studies have documented long-term trends in water concentrations of currently used pesticides with sufficient consistency in locations, timing, and methods to be conclusive. Recently, however, a major change has occurred in herbicide use patterns for corn and soybeans, with a new compound, acetochlor, partially replacing alachlor beginning in 1994. The increase in acetochlor concentrations and decrease in alachlor concentrations in the White River from 1994 through 1996 illustrate the direct connection between chemical use and concentrations in streams and in the major rivers into which they flow.

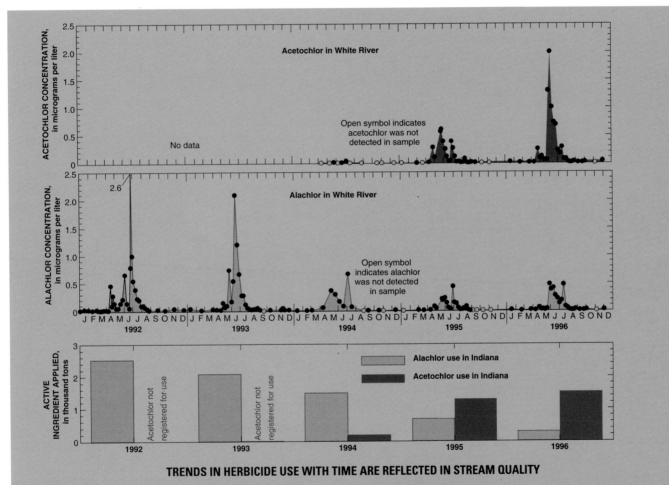

TRENDS IN HERBICIDE USE WITH TIME ARE REFLECTED IN STREAM QUALITY

Alachlor concentrations in streams in the White River steadily declined from 1992 through 1996 and corresponded with a decline in alachlor use in the basin. Application of acetochlor, a corn herbicide registered for use in 1994, has partially replaced the use of alachlor in the basin. Acetochlor was detected at only trace concentrations during the 1994 growing season. By 1996, acetochlor was commonly detected in the White River, where a peak concentration of about 2 µg/L was measured.

STREAMS AND GROUND WATER RESPOND DIFFERENTLY TO CHANGE

Generally, as pesticide use in a basin changes, concentrations in streams quickly reflect these changes. In ground water, however, responses to trends in pesticide-use patterns will be highly variable depending on the nature of the flow system and variability in flow pathways, well depth, and other factors. For the most part, changes in concentrations of pesticides in ground water are much slower than in streams, and responses of ground water to changing use can be delayed for years or decades in some systems.

Study Unit Reports

This report is based on the summary reports of the first 20 NAWQA Study Units, listed below in alphabetical order by Study Unit name. To view an electronic version of a report or to order copies via the World Wide Web, access

http://water.usgs.gov/lookup/get?circXXXX
(where XXXX is the Circular number listed below).

Spruill, T.B., Harned, D.A., Ruhl, P.M., Eimers, J.L., McMahon, G., Smith, K.E., Galeone, D.R., and Woodside, M.D., 1998, Water quality in the **Albemarle–Pamlico Drainage Basin**, North Carolina and Virginia, 1992–95: U.S. Geological Survey Circular 1157, 36 p.

Frick, E.A., Hippe, D.J., Buell, G.R., Couch, C.A., Hopkins, E.H., Wangsness, D.J., Garrett, J.W., 1998, Water quality in the **Apalachicola–Chattahoochee–Flint River Basin**, Georgia, Alabama, and Florida, 1992–95: U.S. Geological Survey Circular 1164, 38 p.

Williamson, A.K., Munn, M.D., Ryker, S.J., Wagner, R.J., Ebbert, J.C., and Vanderpool, A.M., 1998, Water quality in the **Central Columbia Plateau**, Washington and Idaho, 1992–95: U.S. Geological Survey Circular 1144, 35 p.

Frenzel, S.A., Swanson, R.B., Huntzinger, T.L., Stamer, J.K., Emmons, P.J., and Zelt, R.B., 1998, Water quality in the **Central Nebraska Basins**, Nebraska, 1992–95: U.S. Geological Survey Circular 1163, 33 p.

Garabedian, S.P., Coles, J.F., Grady, S.J., Trench, E.C.T., and Zimmerman, M.J., 1998, Water quality in the **Connecticut, Housatonic, and Thames River Basins**, Connecticut, Massachusetts, New Hampshire, New York, and Vermont, 1992–95: U.S. Geological Survey Circular 1155, 32 p.

Berndt, M.P., Hatzell, H.H., Crandall, C.A., Turtora, M., Pittman, J.R., and Oaksford, E.T., 1998, Water quality in the **Georgia–Florida Coastal Plain**, Georgia and Florida, 1992–96: U.S. Geological Survey Circular 1151, 34 p.

Wall, G.R., Riva-Murray, K., and Phillips, P.J., 1998, Water quality in the **Hudson River Basin**, New York and adjacent states, 1992–95: U.S. Geological Survey Circular 1165, 32 p.

Lindsey, B.D., Breen, K.J., Bilger, M.D., and Brightbill, R.A., 1998, Water quality in the **Lower Susquehanna River Basin**, Pennsylvania and Maryland, 1992–95: U.S. Geological Survey Circular 1168, 38 p.

Bevans, H.E., Lico, M.S., and Lawrence, S.J., 1998, Water quality in the Las Vegas Valley area and the Carson and Truckee River Basins (**Nevada Basin and Range**), Nevada and California, 1992–96: U.S. Geological Survey Circular 1170, 47 p.

Petersen, J.C., Adamski, J.C., Bell, R.W., Davis, J.V., Femmer, S.R., Freiwald, D.A., and Joseph, R.L., 1998, Water quality in the **Ozark Plateaus**, Arkansas, Kansas, Missouri, and Oklahoma, 1992–95: U.S. Geological Survey Circular 1158, 33 p.

Ator, S.W., Blomquist, J.D., Brakebill, J.W., Denis, J.M., Ferrari, M.J., Miller, C.V., and Zappia, H., 1998, Water quality in the **Potomac River Basin**, Maryland, Pennsylvania, Virginia, West Virginia, and the District of Columbia, 1992–96: U.S. Geological Survey Circular 1166, 38 p.

Stoner, J.D., Lorenz, D.L., Goldstein, R.M., Brigham, M.E., and Cowdery, T.K., 1998, Water quality in the **Red River of the North Basin**, Minnesota, North Dakota, and South Dakota, 1992–95: U.S. Geological Circular 1169, 33 p.

Levings, G.W., Healy, D.F., Richey, S.F., and Carter, L.F., 1998, Water quality in the **Rio Grande Valley**, Colorado, New Mexico, and Texas, 1992–95: U.S. Geological Survey Circular 1162, 39 p.

Dubrovsky, N.M., Kratzer, C.R., Brown, L.R., Gronberg, J.M., and Burow, K.R., 1998, Water quality in the **San Joaquin–Tulare Basins**, California, 1992–95: U.S. Geological Survey Circular 1159, 38 p.

Dennehy, K.F., Litke, D.W., Tate, C.M., Qi, S.L., McMahon, P.B., Bruce, B.W., Kimbrough, R.A., and Heiny, J.S., 1998, Water quality in the **South Platte River Basin**, Colorado, Nebraska, and Wyoming, 1992–95: U.S. Geological Survey Circular 1167, 38 p.

Land, L.F., Moring, J.B., Van Metre, P.C., Reutter, D.C., Mahler, B.J., Shipp, A.A., and Ulery, R.L., 1998, Water quality in the **Trinity River Basin**, Texas, 1992–95: U.S. Geological Survey Circular 1171, 39 p.

Clark, G.M., Maret, T.R., Rupert, M.G., Maupin, M.A., Low, W.H., and Ott, D.S., 1998, Water quality in the **Upper Snake River Basin**, Idaho and Wyoming, 1992–95: U.S. Geological Survey Circular 1160, 35 p.

Peters, C.A., Robertson, D.M., Saad, D.A., Sullivan, D.J., Scudder, B.C., Fitzpatrick, F.A., Richards, K.D., Stewart, J.S., Fitzgerald, S.A., and Lenz, B.N., 1998, Water quality in the **Western Lake Michigan Drainages**, Wisconsin and Michigan, 1992–95: U.S. Geological Survey Circular 1156, 40 p.

Fenelon, J.M., 1998, Water quality in the **White River Basin**, Indiana, 1992–96: U.S. Geological Survey Circular 1150, 34 p.

Wentz, D.A., Bonn, B.A., Carpenter, K.D., Hinkle, S.R., Janet, M.L., Rinella, F.A., Uhrich, M.A., Waite, I.R., Laenen, A., and Bencala, K.E., 1998, Water quality in the **Willamette Basin**, Oregon, 1991–95: U.S. Geological Survey Circular 1161, 34 p.

References Cited

1. Knopman, D.S., and Smith, R.A., 1993, Twenty years of the Clean Water Act: Environment, v. 35, no. 1, p. 17–34.

2. Hitt, K.J., 1994, Refining 1970's land-use data with 1990 population data to indicate new residential development: U.S. Geological Survey Water-Resources Investigations Report 94–4250, 15 p.

3. Bachman, L.J., Lindsey, B.D., Brakebill, J., and Powars, D.S., 1998, Ground-water discharge and base flow nitrate loads of nontidal streams, and their relation to a hydrogeomorphic classification of the Chesapeake Bay Watershed, Middle Atlantic Coast: U.S. Geological Survey Water-Resources Investigations Report 98–4059, 71 p.

4. Fisher, D.C., and Oppenheimer, M., 1991, Atmospheric nitrogen deposition and the Chesapeake Bay Estuary: AMBIO, v. 20, no. 3–4, p. 102–108.

5. Gilliom, R.J., Alley, W.M., and Gurtz, M.E., 1995, Design of the National Water-Quality Assessment Program—Occurrence and distribution of water-quality conditions: U.S. Geological Survey Circular 1112, 33 p.

6. Barbash, J.E., and Resek, E.A., 1996, Pesticides in ground water—Distribution, trends, and governing factors, v. 2 *of* Gilliom, R.J., ed., Pesticides in the hydrologic system: Chelsea, Mich., Ann Arbor Press, 588 p.

7. Zynjuk, L.D., and Majedi, B.F., 1996, January 1996 floods deliver large loads of nutrients and sediment to the Chesapeake Bay: U.S. Geological Survey Fact Sheet 140–96, 2 p.

8. U.S. Environmental Protection Agency and others, 1998, Clean Water Action Plan—Restoring and protecting America's waters: U.S. Environmental Protection Agency report EPA–840–R–98–001, 89 p.

9. Speiran, G.K., 1996, Geohydrology and geochemistry near coastal ground-water-discharge areas of the Eastern Shore, Virginia: U.S. Geological Survey Water-Supply Paper 2479, 73 p.

10. Tesoriero, A.J., and Voss, F.D., 1997, Predicting the probability of elevated nitrate concentrations in the Puget Sound Basin: Ground Water, v. 35, p. 1029–1039.

11. Litke, D.W., 1999, A review of phosphorus control measures in the United States and their effects on water quality: U.S. Geological Survey Water-Resources Investigations Report 99–4007, 38 p.

12. Puckett, L.J., 1994, Nonpoint and point sources of nitrogen in major watersheds of the United States: U.S. Geological Survey Water-Resources Investigations Report 94–4001, 9 p.

13. Andrilenas, P.A., 1974, Farmers' use of pesticides in 1971—Quantities: U.S. Department of Agriculture, Economic Research Service, Agricultural Economic Report No. 252, 56 p.

14. Eichers, T., Andrilenas, P., Jenkins, R., and Fox, A., 1968, Quantities of pesticides used by farmers in 1964: U.S. Department of Agriculture, Economic Research Service, Agricultural Economic Report No. 131, 37 p.

15. Eichers, T., Andrilenas, P., Blake, H., Jenkins, R., and Fox, A., 1970, Quantities of pesticides used by farmers in 1966: U.S. Department of Agriculture, Economic Research Service, Agricultural Economic Report No. 179, 61 p.

16. Eichers, T.R., Andrilenas, P.A., and Anderson, T.W., 1978, Farmers' use of pesticides in 1976: U.S. Department of Agriculture, Economics, Statistics, and Cooperative Service, Agricultural Economic Report No. 418, 58 p.

17. Aspelin, A.L., 1997, Pesticides industry sales and usage, 1994 and 1995 market estimates: U.S. Environmental Protection Agency, Washington, DC, 733–R–97–002, 35 p.

18. Shelton, L.R., 1994, Field guide for collecting and processing stream-water samples for the National Water-Quality Assessment Program: U.S. Geological Survey Open-File Report 94–455, 42 p.

19. Shelton, L.R., and Capel, P.D., 1994, Guidelines for collecting and processing samples of stream bed sediment for analysis of trace elements and organic contaminants for the National Water-Quality Assessment Program: U.S. Geological Survey Open-File Report 94–458, 20 p.

20. Crawford, J.K., and Luoma, S.N., 1994, Guidelines for studies of contaminants in biological tissues for the National Water-Quality Assessment Program: U.S Geological Survey Open-File Report 92–494, 69 p.

21. Lapham, W.W., Wilde, F.D., and Koterba, M.T., 1995, Ground-water data-collection protocols and procedures for the National Water-Quality Assessment Program—Selection, installation, and documentation of wells, and collection of related data: U.S. Geological Survey Open-File Report 95–398, 69 p.

22. Koterba, M.T., Wilde, F.D., and Lapham, W.W., 1995, Ground-water data-collection protocols and procedures for the National Water-Quality Assessment Program—Collection and documentation of water-quality samples and related data: U.S. Geological Survey Open-File Report 95–399, 113 p.

23. Mueller, D.K., and Stoner, J.D., 1998, Identifying the potential for nitrate contamination of streams in agricultural areas of the United States, *in* Proceedings of the National Water Quality Monitoring Conference, July 7–9, 1998, Reno, Nev., p. III163–III173.

24. National Atmospheric Deposition Program, 1997, National atmospheric deposition program (NRSP-3)/ National trends network, 1997: NADP/NTN Program Office, Illinois State Water Survey, Champaign, Ill.

25. Gilliom, R.J., Mueller, D.K., and Nowell, L.H., 1998, Methods for comparing water-quality conditions among assessment study units, 1992–1995: U.S. Geological Survey Open-File Report 97–589, 54 p.

26. Mueller, D.K., Hamilton, P.A., Helsel, D.R., Hitt, K.J., and Ruddy, B.C., 1995, Nutrients in ground water and surface water of the United States—An analysis of data through 1992: U.S. Geological Survey Water-Resources Investigations Report 95–4031, 74 p.

27. Mueller, D.K., and Helsel, D.R., 1996, Nutrients in the Nation's waters—Too much of a good thing?: U.S. Geological Survey Circular 1136, 24 p.

28. U.S. Environmental Protection Agency, 1996, Drinking water regulations and health advisories: U.S. Environmental Protection Agency report EPA 822–B–96–001 [variously paged].

29. U.S. Environmental Protection Agency, 1986, Quality criteria for water—1986: U.S. Environmental Protection Agency report EPA 440/5–86–001 [variously paged].

30. Nolan, B.T., Ruddy, B.C., Hitt, K.J., and Helsel, D.R., 1997, Risk of nitrate in ground waters of the United States—A national perspective: Environmental Science & Technology, v. 31, no. 8, p. 2229–2236.

31. Nowell, L.H., Capel, P.D., and Dileanis, P.D., in press, Pesticides in stream sediment and aquatic biota—Distribution, trends, and governing factors, v. 4 *of* Gilliom, R.J., ed., Pesticides in the hydrologic system: Chelsea, Mich., Ann Arbor Press.

32. Carson, R., 1962, Silent Spring: Boston, Houghton Mifflin; Cambridge, Mass., Riverside Press.

33. Goodbred, S.L., Gilliom, R.J., Gross, T.S., Denslow, N.P., Bryant, W.L., and Schoeb, T.R., 1997, Reconnaissance of 17β-estradiol, 11-ketotestosterone, vitellogenin, and gonad histopathology in common carp of United States Streams—Potential for contaminant-induced endocrine disruption: U.S. Geological Survey Open-File Report 96–627, 47 p.

34. Kolpin, D.W., Thurman, E.M., and Linhart, S.M., 1998, The environmental occurrence of degradates in ground water: Archives of Environmental Contamination and Toxicology, v. 35, p. 385–390.

35. U.S. Environmental Protection Agency, 1990, Suspended, cancelled, and restricted pesticides: U.S. Environmental Protection Agency, Office of Pesticides and Toxic Substances 20T–4002.

36. Buell, G.R., and Couch, C.A., 1995, National Water Quality Assessment Program—Environmental distribution of organochlorine compounds in the Apalachicola-Chattahoochee-Flint River Basin, *in* Hatcher, K.J., ed., Proceedings of the 1995 Georgia Water Resources Conference: Athens, Ga., Vinson Institute of Government, University of Georgia, p. 46–53.

37. Meade, R.H., ed., 1995, Contaminants in the Mississippi River, 1987–92: U.S. Geological Survey Circular 1133, 139 p.

38. Stamer, J.K., and Wieczoreck, M.E., 1995, Pesticides in streams in Central Nebraska: U.S. Geological Survey Fact Sheet FS-232-95, 4 p.

39. U.S. Environmental Protection Agency, 1997, The incidence and severity of sediment contamination in surface waters of the United States, v. 1 of National sediment quality survey: U.S. Environmental Protection Agency, Office of Science and Technology, EPA 823–R–97–006.

40. Newell, A.J., Johnson, D.W., and Allen, L.K., 1987, Niagara River Biota Contamination Project—Fish flesh criteria for piscivorous wildlife: New York State Department of Environmental Conservation, Division of Fish and Wildlife, Bureau of Environmental Protection, Technical Report 87–3.

41. U.S. Environmental Protection Agency, 1995, Guidance for assessing chemical contaminant data for use in fish advisories, v. 1 of Fish sampling and analysis (2d ed.): U.S. Environmental Protection Agency, Office of Water, EPA 823–R–95–007.

42. Rupert, M.G., 1998, Probability of detecting atrazine/desethylatrazine and elevated concentrations of nitrate (NO_2+NO_3-N) in ground water in the Idaho part of the Upper Snake River Basin: U.S. Geological Survey Water-Resources Investigation Report 98–4203, 32 p.

43. Larson, S.J., Capel, P.D., and Majewski, M.S., 1997, Pesticides in surface waters—Distribution, trends, and governing factors, v. 3 *of* Gilliom, R.J., ed., Pesticides in the hydrologic system: Chelsea, Mich., Ann Arbor Press, 373 p.

44. Scribner, E.A., Goolsby, D.A., Thurman, E.M., and Battaglin, W.A., 1998, Reconnaissance for selected herbicides, metabolites, and nutrients in streams of nine Midwestern states: U.S. Geological Survey Open-File Report 98–181, 36 p.

45. VanMetre, P.C., Wilson, J.T., Callender, E., and Fuller, C.C., 1998, Similar rates of decrease of persistent, hydrophobic and particle-reactive contaminants in riverine systems: Environmental Science & Technology, v. 32, no. 21, p. 3312–3317.

46. VanMetre, P.C., and Callender, E., in press, Trends in sediment quality in response to urbanization, *in* Buxton, H., and Morganwalp, D.W., eds., Toxic Substances Hydrology Program, Proceedings of the Technical Meeting, March 8–12, 1999, Charleston, S.C.

47. U.S. Fish and Wildlife Service, 1992, U.S. Fish and Wildlife Service National Contaminant Biomonitoring Program Fish Data File, 1969–1986: Fish and Wildlife Service National Fisheries Contaminant Research Center, 4200 New Haven Road, Columbia, Mo. 65201, Lotus and ASCII files (8/24/92).

48. Rapaport, R.A., Urban, N.R., Capel, P.D., Baker, J.E., Looney, B.B., Eisenreich, S.J., and Gorham, E., 1985, "New" DDT inputs to North America—Atmospheric deposition: Chemosphere, v. 14, no. 9, p. 1167–1173.

49. U.S. Environmental Protection Agency, 1983, Analysis of the risks and benefits of seven chemicals used for subterranean termite control: U.S. Environmental Protection Agency, Office of Pesticides and Toxic Substances, Office of Pesticide Programs EPA–540/9–83–005 [variously paged].

50. Esworthy, R.F., 1987, Incremental benefit analysis—Restricted use of all pesticides registered for subterranean termite control: U.S. Environmental Protection Agency, Benefits and Use Division, Economic Analysis Branch.